MW00463383

Thinking about Consciousness

Consciousness is widely regarded as an intractable mystery. Many scientists and philosophers view it as an enigma whose solution waits on some unforeseeable theoretical breakthrough. David Papineau argues that this pessimism is quite misplaced. Consciousness seems mysterious, not because of any hidden essence, but only because we humans think about it in a special way. *Thinking about Consciousness* analyses this special mode of thought in detail, and exposes the ways in which it can lead us into confusions about consciousness.

At the heart of the book lies a distinction between two ways of thinking about conscious states. We humans can think about conscious states materially, as normal items inhabiting the material world. But we can also think about them phenomenally, as items that feel a certain way. Dualists hold that this phenomenal mode of thought describes some special non-material reality. But David Papineau argues that it is invalid to move from a distinctive phenomenal mode of thought to a distinct non-material reality. By carefully analysing the structure of phenomenal concepts, he is able to expose the flaws in the standard arguments for dualism, while at the same time explaining why dualism can seem so intuitively compelling.

Thinking about Consciousness also casts a new light on contemporary scientific research into consciousness. Much of this research is motivated by the apparently 'hard problem' of identifying the referents of phenomenal concepts. David Papineau argues that such research promises less than it can deliver. Once phenomenal concepts are recognised for what they are, many of the questions posed by consciousness research turn out to be irredeemably vague.

This is the first book to provide a detailed analysis of phenomenal concepts from a materialist point of view. By recognising the importance of phenomenal thinking, David Papineau is able to place a materialist account of consciousness on a firm foundation, and to lay many traditional problems of consciousness to rest.

Thinking about
Consciousness

DAVID PAPINEAU

CLARENDON PRESS · OXFORD

This book has been printed digitally and produced in a standard specification in order to ensure its continuing availability

OXFORD
UNIVERSITY PRESS

Great Clarendon Street, Oxford OX2 6DP

Oxford University Press is a department of the University of Oxford.
It furthers the University's objective of excellence in research, scholarship,
and education by publishing worldwide in

Oxford New York

Auckland Bangkok Buenos Aires Cape Town Chennai
Dar es Salaam Delhi Hong Kong Istanbul Karachi Kolkata
Kuala Lumpur Madrid Melbourne Mexico City Mumbai Nairobi
São Paulo Shanghai Taipei Tokyo Toronto

Oxford is a registered trade mark of Oxford University Press
in the UK and in certain other countries

Published in the United States
by Oxford University Press Inc., New York

Reprinted 2004

ISBN 0-19-924382-4

For Louis

Preface

I seem to have been writing this book for some time. A while ago I formed the plan of juxtaposing a number of already written pieces to form a book on consciousness. But in the course of tidying and clarifying, my views kept developing and expanding, and now little of the original material is left, and much has been added. For what it is worth, parts of Chapter 1 and most of the Appendix descend from 'The Rise of Physicalism', first published in 2000, but written rather earlier. Parts of Chapters 2 and 6 can be traced back to my first attacks on consciousness in *Philosophical Naturalism* (1993*a*) and in 'Physicalism, Consciousness and the Antipathetic Fallacy' (1993*b*). Chapters 3 and 4 are pretty much new. Chapter 5 has affinities with 'Mind the Gap' (1998). Chapter 7 started as 'Theories of Consciousness' (2001), but no longer bears much relation to that article.

In the course of writing this book I have had many opportunities to try out my views. Particularly helpful have been a number of occasions that allowed extended discussion. In the Autumn Term of 1999 I held a research seminar on consciousness at King's College London, and was greatly helped by the graduate students, colleagues, and vistors who attended. I can particularly remember comments from Heather Gert, Matteo Mameli, Nick Shea, Finn Spicer, and Scott Sturgeon. In the Spring of 2000 I was invited to the New York University Research Seminar on Consciousness, where an early draft of parts of the book was subject to the scrutiny of Ned Block and Tom Nagel. In March 2001 I conducted a week-long 'superseminar' at the University of Kansas, hosted by Sarah Sawyer and Jack Bricke. I went to the University of Bielefeld for two days in May 2001, where my commentators included Ansgar Beckermann, Martin Carrier, Andreas Hüttemann, and Christian Nimtz.

Apart from these occasions, I have been invited to speak at conferences on consciousness in Oxford in 1997, Bremen in 1998, King's College London in 1999, and Nottingham in 2000. I learnt much at all these meetings, and would like to thank the organizers for inviting me. I am also grateful to all those who made comments on other talks I have given on consciousness and related topics over the past few years in Durban, Hamburg, Bradford, Bogota, Lisbon, Dublin, Budapest, Athens, Nottingham, Oxford, Hertfordshire, Reading, Cardiff, New York, Durham, Middlesex, and various venues in London.

A number of people have been kind enough to read drafts of the book, or parts of it, and give me written comments. I am very grateful indeed to Peter Carruthers, Peter Goldie, Keith Hossack, Sarah Sawyer, Gabriel Segal, and especially Scott Sturgeon.

Many other individuals apart from those already mentioned have helped me. I know that I will have forgotten some people I shouldn't have, and I can only ask them to forgive my memory. I can remember assistance from Kati Balogh, David Chalmers, John Cottingham, Tim Crane, Jerome Dokic, Ray Dolan, Kati Farkas, Chris Frith, Christopher Hill, Jim Hopkins, Tony Jack, Bob Kirk, Joe Levine, Brian Loar, Barry Loewer, Guy Longworth, Jonathan Lowe, Graham MacDonald, Brian McLaughlin, Mike Martin, Barbara Montero, Martine Nida-Rümelin, Lucy O'Brien, Anthony O'Hear, Michael Pauen, Stathis Psillos, Diana Raffman, Georges Rey, Alejandro Rojas, Mark Sainsbury, Maja Spener, David Spurrett, Joan Steigerwald, Michael Tye, and Antonio Zilhao.

Various further thanks are due. I am indebted to the British Academy for a Research Leave Award in 1998, to King's College London for allowing me a sabbatical term in 1999, and to the Leverhulme Foundation for a Research Fellowship in 1999–2000. I would also like to thank Peter Momtchiloff of Oxford University Press for all his encouragement and advice, and Jean van Altena for her intelligent and helpful copy-editing. I was extremely glad when Wes Schoch gave me permission to use his photograph of the 'reefs, sandbanks, waves and rock pools' at Isipingo (p. 107) as the cover illustration. Finally, I am very grateful indeed to all my colleagues in the Philosophy Department at King's for making it such an extremely congenial place to work.

Contents

INTRODUCTION

1 *Mystery—What Mystery?*

Consciousness is widely regarded as an intractable mystery. As soon as we start thinking about it, we find ourselves pulled in two quite opposite directions, and there can seem no good way of resolving the conflict.

On the one hand, it seems clear that consciousness must be a normal part of the material world. Conscious states clearly affect our bodily movements. But surely anything that so produces material effects must itself be a material state.

On the other hand, it seems absurd to identify conscious states with material states. Conscious states involve awareness, feelings, the subjectivity of experience. How could mere matter on its own account for the miracle of subjective feelings?

In the face of this dilemma, many contemporary thinkers counsel despair. They conclude that we lack the intellectual wherewithal to understand consciousness.

Some suggest that this failing may be temporary. Even if our present science is inadequate, they hope that the concepts of some future theory will show us how to unlock the puzzle of consciousness. Others are more pessimistic, and fear that the human mind is limited in ways that will permanently bar us from understanding the mystery.

For myself, I think that all this gloom is quite misplaced. We don't need any fancy new concepts to understand consciousness. For there isn't anything really mysterious about it in the first place.

The basic puzzle, as I presented it above, was to reconcile the causal efficacy of mental states with their subjectivity. Well, I agree entirely with the thought that, in order for conscious states to be causally efficacious, they must be material states.[1] But I don't see why this should leave us with a puzzle about subjective feelings. Why not just accept that having a subjective feeling is being in a material state? What would you expect it to feel like to be in that material state? Like nothing? Why? *That's* what it is like to be in that material state.

2 *The Intuition of Distinctness*

I recognize, though, that there certainly *seems* to be a mystery here. But I don't think that this is because there is something unfathomable about the thesis that conscious states are material. Rather, it is because something prevents us from ever fully accepting this thesis in the first place, and convinces us that conscious states are *not* material states. And then, of course, everything does seem mysterious.

For, as soon as you suppose that conscious states *are* distinct from material states, then some very puzzling questions become unavoidable. How can these extra conscious states possibly exert any causal influence on the material realm? And why are they there at all? By what mysterious power do our material brains generate these additional conscious feelings?

Note, however, that these puzzles arise only because of the initial dualist separation of mind from brain. They would simply dissolve if we fully accepted that conscious states are one and the same as brain states. For, if we really believed this, then we could simply view conscious causes as operating in the same way as other material causes. Nor would there be any puzzle about brain states 'generating' extra non-material feelings. If feelings are one and the same as brain

[1] As we shall see, the important point here isn't the traditional worry that a non-material mind and a material brain would be qualitatively too *different* to enter into any causal intercourse. Nothing in what I say will rest on this thought. Rather, the real issue is 'the causal completeness of physics'—that is, the thesis that all physical effects already have *physical* causes. This seems to leave no room for non-material mental causes to make a difference to material effects.

states, then brain states don't 'generate' a further realm of feelings (or 'give rise to' them, or 'accompany' them, or 'are correlated with' them). Rather, the brain states *are* the feelings. They are what they are, and couldn't be otherwise.[2]

Still, as I said, it is very hard for us properly to accept that conscious feelings are nothing but material states. Something stops us embracing such identities. We find it almost impossible to free ourselves from the dualist thought that conscious feelings must be something *additional* to any material goings-on. And then, once more, we are stuck with the intractable philosophical puzzles.

This book is an attempt to understand this dualist compulsion, and free us from its grip. A successful materialism must explain the compelling intuition that the mind is ontologically distinct from the material world. This anti-materialist intuition comes so naturally to us that we are unlikely to become persuaded of materialism simply by *arguments*. We can rehearse the considerations in favour, and show that the counter-arguments are not compelling. But as long as the contrary intuition remains, this all seems like a trick. There must be a flaw in the argument, we feel, because it is *obvious* that conscious states are not material states.

So a successful materialism must identify the source of this contrary intuition. It needs to explain why materialism should seem so obviously false, if it is indeed true.

3 *A Need for Therapy*

Wittgenstein thought that all philosophy should be therapy. In his view, philosophical problems arise because we allow superficial features of our thinking to seduce us into confusions. The appropriate cure, Wittgenstein thought, is to become sensitized to the deeper

[2] Some readers may be uneasy with my implicit assumption that materialism will equate conscious states specifically with *brain* states. Might not consciousness depend on material matters outside the brain, or even outside the body, as well as within? I find this suggestion strange, and shall continue to assume that only brain states matter to consciousness. But most of the arguments which follow do not depend on this assumption. I shall make some further comments on this issue in Chapters 1 and 7.

structure of our conceptual framework. This philosophical therapy will then free us from muddled thinking.

I reject this conception of philosophy almost entirely. I hold that, on the contrary, nearly all important philosophical problems are occasioned by real tensions in our overall theories of the world, and that their resolution therefore calls for substantial theoretical advances, rather than mere conceptual tidying.

Still, when it comes to the particular topic of consciousness, I think Wittgenstein was right. Here our problems are conceptual rather than theoretical. The difficulty isn't that our overall theories articulate inconsistent claims about consciousness. Rather, we get tangled up before we even start theorizing. We get confused by superficial features of our thinking, in the way Wittgenstein had in mind. This happens because we have a special set of concepts for thinking about conscious states, and the structure of these concepts can easily lead us astray. To resolve our philosophical difficulties, we need first to understand this special conceptual structure.

In line with this diagnosis, I shall not be offering any '*theory* of consciousness' in these pages. There are many such theories on offer nowadays, from both scientists and philosophers, and I shall make some comments on the prospects for such theorizing in Chapter 7. But, in my view, such theorizing is premature. The first step is to unravel our confusions. Then there may be room for 'theories of consciousness' (though Chapter 7 explains why I have my doubts). The first task, however, is to clear away the conceptual tangles. To make progress with consciousness, we need therapy, not theories.

4 *Ontological Monism, Conceptual Dualism*

The main body of this book, Chapters 2–6, aims to offer just this kind of therapy. I seek to understand the source of our bewilderment about consciousness, and thereby free us from its grip.

The key is to recognize that, even if conscious states *are* material states at the ontological level, we have two different ways of thinking about these states at the conceptual level. As well as thinking of them *as* material states, we can also think of them *as* feelings, by using special 'phenomenal concepts'. By carefully analysing the workings of these phenomenal concepts, I am able to explain why it should seem so obvious that conscious states are distinct from material states, even though in reality they are not.

Hence the title of this book—*Thinking about Consciousness*. This isn't just a book about consciousness. It is more specifically a book about the special ways in which we *think* about consciousness. Such self-conscious reflexivity isn't always a good strategy for intellectual progress, and indeed is often positively unhelpful. But it is just what we need for the peculiar topic of consciousness.

The general line adopted in this book is no longer new. Plenty of materialist philosophers of consciousness now combine the *ontologically monist* view that conscious states are material states with the *conceptually dualist* doctrine that we have two distinct sets of concepts for thinking about these states, including a special set of phenomenal concepts. (Cf. Peacocke 1989, Loar 1990, Papineau 1993*a*, 1993*b*; Sturgeon 1994, Hill 1997, Hill and McLaughlin 1998, Tye 1999.) Indeed, this conceptual dualism is quickly becoming the orthodoxy among analytic philosophers who defend a materialist view of consciousness.

Still, this book is intended to go beyond this emerging consensus in two ways. First, I offer a detailed account of the working of phenomenal concepts. Most materialist philosophers are interested in phenomenal concepts only because they can use these concepts to block standard anti-materialist arguments, such as Jackson's knowledge argument, Kripke's modal argument, and Levine's argument from 'the explanatory gap' (Jackson 1982, 1986, Kripke 1971, 1972, 1980, Levine 1983). Because of this, they tend not to dwell on the nature of these phenomenal concepts, apart perhaps from making some general suggestions about their dependence on imagination, or their similarity to indexical constructions. By contrast, I analyse the workings of these concepts in great detail, explaining exactly how they relate to other mental powers, and in what respects they do and do not resemble indexicals.

Second, I go beyond other contemporary materialists in offering an explicit account of why materialism should be so hard to believe, if it is true. As I said above, a successful materialism needs to diagnose and cure this intuitive antipathy to materialism, otherwise materialism will seem impossible to believe, even after all the arguments are done. But the intuitive pull of dualism has not received the attention it deserves in the current literature. To the extent that materialist philosophers have addressed it, they have tended to assume that the attraction of dualism simply derives from one or another of the standard anti-materialist arguments, like Jackson's or Kripke's or Levine's.

I argue in what follows that this diagnosis is mistaken. Let me clarify the precise point at issue here. It is not whether the standard anti-materialist arguments succeed in disproving materialism. Along with other materialists, I think they do not, and explain why when I discuss them. The issue, rather, is whether, even given their unsoundness, the standard anti-materialist arguments can nevertheless account for the widespread *conviction* that materialism is false. Perhaps, despite their unsoundness, they are still plausible enough to seduce the unsophisticated into dualism.

I argue that the standard anti-materialist arguments do not do even this much. In order to show this, I point out that each of these arguments appeals to some feature of our thinking about conscious states that is also found in our thinking about other subject areas. Yet we do not find corresponding intuitions of ontological distinctness in these other subject areas. I conclude that the persistent intuition of mind–brain distinctness is due to some further feature of the way we think about conscious states, beyond the features appealed to in the standard anti-materialist arguments.

5 *Understanding the Intuition of Distinctness*

I have a theory about this special feature. I hold that the intuition of distinctness stems from the peculiar way in which phenomenal concepts of conscious states standardly *exemplify* or *simulate* versions of those conscious states themselves. This can sow great confusion when we come to contrast this phenomenal way of thinking about conscious states with other ways of thinking about them, and in particular with thinking of them as material states. Since the latter, non-phenomenal modes of thought do *not* similarly exemplify or simulate conscious states, we feel that they 'leave out' the feelings themselves. And so we conclude that the feelings themselves must be something different from the material states we think about non-phenomenally.

If we stop to think about this line of reasoning, we can see that it is fallacious. In previous writings I have dubbed it the 'antipathetic fallacy' (Papineau 1993*a*, 1993*b*, 1995). It involves a kind of use–mention fallacy. That material modes of thought don't *activate* feelings doesn't mean they can't *refer* to feelings. So this line of reasoning gives us no real cause to distrust materialism. But, for all that, it is terribly seductive. It is ubiquitous in everyday discussions of consciousness, and the reason, I am convinced, why so many people find materialism so difficult to believe.

Thus consider the standard rhetorical ploy used against materialism. 'How can technicolour phenomenology arise from soggy grey matter?' (McGinn 1991). Here we are first invited to activate a version of the experience of colour (think of *what it is like* to see technicolour reds and greens). Then we are invited to think non-phenomenally about the putative material equivalent of colour experience (think about a section of squishy brain tissue). Now, we don't of course *activate* anything like colour experiences in the latter case, when we think about brains. But that doesn't mean we aren't thinking *about* colour experiences when we do so. In general, thinking about something doesn't require activating some version of it.

The way to free ourselves from the seductive fallacy is to understand the special structure of our phenomenal concepts. We need to recognize the existence of these concepts, and to note in particular how they *simulate* the feelings they refer to. Then we can see why it is so natural to conclude that other, non-phenomenal concepts inevitably 'leave out' the conscious feelings. And we can also see that, while there is a sense in which this conclusion is true (the non-phenomenal concepts don't *use* the feelings), this is not inconsistent with materialism (for the non-phenomenal concepts may still *refer to* the feelings).

6 *The Details of Materialism*

All this, as I said, comes in the main body of the book, in Chapters 2–6. Chapter 1 is devoted to a rather different set of issues. Here I look at the rationale for embracing a materialist view of consciousness in the first place. I don't take materialism to be obvious, or some kind of default position which we should automatically embrace if only we can remove the barriers to its acceptance. On the contrary, I regard it as a rather eccentric position, which stands in need of serious argumentative support. (Certainly it is a minority attitude from a historical point of view. Few philosophers or scientists have been materialists about consciousness until relatively recently, for reasons I shall mention in a moment.)

So materialism stands in need of an argument. However, such an argument is not hard to find. Recall the causal argument alluded to at the beginning of this Introduction. Conscious states clearly affect our bodily movements. But surely anything that so produces a material effect must itself be a material state.

In Chapter 1 I look at this argument in some detail. I lay out its premisses explicitly, and consider how far it is feasible for anti-materialists to deny them.

In some ways I would have preferred to skip this initial chapter. The issues surrounding the causal argument have been explored extensively by recent philosophers, and I do not take myself to have anything especially new to add to this debate. Indeed, at one time I hoped to take the causal argument as read, and start straight off with my analysis of phenomenal concepts.

But it soon became clear to me that this was not really feasible. Anybody writing seriously about mind–brain issues nowadays needs to explain whether they think of materialism in terms of type identity, token identity, realization, or supervenience. They need to explain whether they think of causation in terms of events, facts, or states of affairs. And they need to explain exactly how they understand all these terms, not to mention how they understand the terms which frame the debate in the first place, such as 'material' and 'physical'.

I go through all this in Chapter 1. If you are prepared to take my line on these matters on trust, I would be more than happy for you to jump straight to Chapter 2. But for those who want to be clear about the precise way I am construing materialism, Chapter 1 is the place to look.

One specific issue that arises in chapter 1 is worth mentioning. A crucial premiss in the causal argument—the 'completeness' (or 'causal closure') of physics—turns out to be a relatively recent scientific discovery. The evidence in favour of this premiss has accumulated only over the last century or so. Correspondingly, this premiss was widely disbelieved in the seventeenth, eighteenth, and nineteenth centuries, by serious physical scientists as much as others, which is why, if you ask me, materialism was so little believed until recently.

There is of course no reason why this recent provenance of the completeness of physics should present a problem for materialism. A recently discovered truth is still a truth, and we will still do well to believe its consequences. But it is worth focusing on the historical contingency of the completeness of physics, for it does have the virtue of explaining why philosophical materialism is so much a creature of the late twentieth century. Sceptics sometimes suggest that this popularity is essentially a matter of passing fashion. I am able to argue that, on the contrary, the late rise of philosophical materialism is fully explained by the late scientific emergence of the completeness of physics. (Some of the more detailed historical discussion of this issue has been relegated to an Appendix at the end of the book.)

7 *The Plan of the Book*

After the general materialist arguments of Chapter 1, I turn to the analysis of phenomenal concepts. In Chapter 2 I start with Frank Jackson's knowledge argument. Jackson's argument is designed as an argument for ontological dualism. I show that this ontological conclusion does not follow, but that Jackson's line of thought nevertheless provides an effective demonstration of conceptual dualism—that is, of the existence of distinct phenomenal concepts.

In this chapter I also make some initial comments about the nature of these phenomenal concepts.

Chapter 3 begins with Kripke's modal argument against materialism. In the first instance I simply aim to analyse this argument, and to show that there is a way for the materialist to defuse it. But in the course of this analysis a further feature of phenomenal concepts emerges: if materialism is true, then phenomenal concepts must refer directly, and not by invoking any contingent features of their referents.

In Chapter 4 I build on the points already established to develop a detailed account of phenomenal concepts. I compare phenomenal concepts, which refer to experiences, with perceptual concepts, which standardly refer to observable features of the non-mental world. And I argue that phenomenal concepts paradigmatically draw on exercises of perceptual concepts, in a quotational manner. At the end of this chapter I use this account to cast some light on the ways in which we are immune to error about our own conscious states.

Chapter 5 is concerned with the 'explanatory gap'. I make the following points. Mind–brain identities are indeed inexplicable, but so are many other true identities. By contrast, scientific identities are characteristically open to explanation, in a way that mind–brain identities are not. However, this is simply because scientific and mind–brain identity claims have significantly different structures, and not because there is anything wrong with mind–brain identities. In any case, these matters of relative explanatoriness have little to do with the intuitive feeling that there is a brain–mind gap. This has a different source, which has nothing to do with the fact that mind–brain identities don't explain.

In Chapter 6 I focus on the real source of the intuition of mind–brain distinctness. I first show that the standard accounts of this intuition are inadequate. I then appeal to my analysis of phenomenal concepts to explain the intuition, as arising from the 'antipathetic fallacy', in the way outlined above. That is, I point out that phenomenal concepts activate versions of the feelings they refer to. By contrast, non-phenomenal concepts do not so activate any feelings. And then it is all to easy to slide, via the thought that the non-phenomenal concepts 'leave out' the feelings, to the fallacious conclusion that non-phenomenal concepts cannot refer to feelings.

In the final chapter I consider the prospects for substantial scientific research into consciousness—that is, research which seeks to identify the material referents of phenomenal concepts on the basis of empirical evidence. Nowadays there is a great deal of enthusiasm for such research, among psychologists, neurologists, and other cognitive scientists, as well as among philosophers. But I argue that such research is limited in essential ways. There are questions about the referents of phenomenal concepts that it is quite unable to answer.

However, I do not take this to show that there are mysteries of consciousness which somehow lie beyond the reach of science. Rather, the fault lies in our phenomenal concepts themselves. They are irredeemably *vague* in certain dimensions, in ways that preclude there being any fact of the matter about whether octopuses feel phenomenal pain, say, or whether a silicon-based humanoid would have any kind of phenomenal consciousness. I realize that this suggestion will seem counter-intuitive. Moreover, it calls into question the motivations for much current 'consciousness research'. Nevertheless, I think that there is no basis, beyond outmoded metaphysical thinking, for the conviction that facts about phenomenal consciousness must be sharp. And, in so far as the current enthusiasm for 'consciousness research' rests on this conviction, it would be no bad thing for it to be dampened.

Chapter 1
THE CASE FOR MATERIALISM

1.1 *Introduction*

Books on consciousness often begin by distinguishing between different kinds of consciousness. We are told about self-consciousness and sentience, creature consciousness and state consciousness, phenomenal consciousness and access consciousness, perceptual consciousness, higher-order consciousness, and so on. I'd rather leave all this until later. Some of these distinctions will become significant in due course, and will be explained when they are needed. Others will not matter to my discussion.

For the moment, all I want to say is that I am concerned with that aspect of consciousness that makes it so philosophically interesting. Namely, that having a conscious experience is *like something*, in Thomas Nagel's striking phrase (1974). It has become standard to use 'phenomenal' or 'subjective' to focus on this feature of consciousness, and I shall adopt these usages in what follows.

The idea is best introduced by examples rather than definitions. ('If you gotta ask, you're never gonna know.') Compare the difference between having your eyes shut and having them open, or between having your teeth drilled with and without an anaesthetic. When your eyes are open, you have a conscious visual experience, and when your teeth are drilled without an anaesthetic, you have a conscious pain. It is like something for you to have these experiences. It is not like that when you close your eyes, or when the anaesthetic takes effect. What you lose in these latter cases are elements of phenomenal or subjective consciousness.[1] From now on, when I say 'conscious', I shall mean this kind of consciousness.

Much of what follows will be concerned with a particular philosophical puzzle about consciousness: namely, the puzzle of how consciousness relates to the physical world. There are other philosophical puzzles about consciousness, but this seems to me the most immediate. We will be ill placed to understand anything about consciousness if we cannot understand its relation to the physical realm.

The puzzle can be posed simply. On the one hand, there is a strong argument for adopting a materialist view of conscious states, for supposing that conscious states must be *part* of the physical world, that they must be *identical* to brain states, or something similar. Yet, on the other hand, there are also strong arguments (and even stronger intuitions) which suggest that conscious states must be *distinct* from any material states.

I believe that in the end the materialist argument wins. Conscious states are material states. This is not to belittle the anti-materialist arguments and intuitions. They are deep and important. We will not grasp consciousness properly unless we understand how to answer them. Still, I think that careful analysis will show that they are flawed, and that the right solution is to embrace materialism.

I shall begin by putting the materialist argument on the table. It is

[1] Some philosophers assume that 'phenomenal' is meant to contrast with *intentional*, and on this basis hold that much recent discussion of consciousness, especially that surrounding David Chalmers's 'hard problem' (1996), is invalidated by an implicit supposition that subjectivity is independent of intentionality (cf. Eilan 1998). It is perhaps worth emphasizing that I don't intend 'phenomenal' to imply 'non-intentional'. I simply mean it as a non-committal term for subjective 'what-it's-likeness'. Nothing yet rules out the possibility that all, or only, intentional states involve phenomenal consciousness. Moreover, since Chalmers also understands 'phenomenal' in this way, his 'hard problem' of phenomenal consciousness will still arise even if phenomenality is not independent of intentionality.

worth taking some care about this, for there are a number of different defences of materialism on offer in the contemporary literature, and not all of them are equally compelling. However, I think that there is one definitive argument for materialism. I shall call this 'the causal argument', and the burden of this first chapter will be to develop this argument and distinguish it from some less effective defences of materialism.

There is a further reason for laying out the argument for materialism carefully. Many contemporary philosophers harbour grave suspicions about materialism. Thus some philosophers contend that the whole idea of materialism is somehow empty, on the grounds that there is no proper way of characterizing the 'physical' realm. (Crane and Mellor 1990, Crane 1991, Segal 2000). And others suggest that contemporary materialism about the conscious mind rests on nothing but fashion or prejudice, unsupported by serious argument (Burge 1993, Clark 1996).

I intend to show that these attitudes are mistaken. The question of how to define 'physical' in the context of the mind-brain debate does raise a number of interesting points, but there is no great difficulty about pinning down a sense precise enough for the purposes at hand. It will prove easier do this, however, after we have rehearsed the argument for materialism. Accordingly, I shall not worry about the meaning of 'physical' at this stage, but simply begin by outlining the case for materialism. Once we have seen what is at issue, it will become clearer how materialists can best understand the meaning of 'physical', and I shall return to this issue at the end of the chapter.

There is one terminological point which I do need to address at this point, however. When I do fix a meaning for 'physical' at the end of the chapter, I shall read this term in a relatively strict sense, as standing roughly for the kinds of first-order properties studied by the physical sciences. Under the heading of 'materialism', on the other hand, I shall include not only the doctrine that conscious states are identical with physical states in this strict sense, but also the doctrine that they are identical with 'physically realized functional states', or with some other kind of physically realized but not strictly physical states (these possibilities will be explained further in section 1.6

below). It is true that the causal argument can be read as supporting the stricter identification with physical states, and indeed this is how I shall first present it in the next section. But, as we shall see, the causal argument can also be construed as supporting the less strict identification of conscious states with functional or other physically realized states. Since both the strict and the less strict identifications tie conscious states constitutively to the physical world, few of the arguments in this book will require me to decide between them. So it will be useful to have a term which covers both options, and I have adopted 'materialism' for this purpose. Correspondingly, a 'material' state will mean either a physical state in the strict sense or some functional or other physically realized state.

In addition to suspicions about the meaning of 'physical', there is the further allegation mentioned above, that contemporary materialism is nothing but a modish fad. I take the causal argument to be outlined in this chapter to rebut this allegation. The causal argument may not be conclusive, but it certainly shows that the case for materialism goes beyond mere fashion or prejudice.

Some may think that the charge of modishness is supported by historical considerations. Widespread philosophical materialism is a relatively recent phenomenon, largely a creature of the late twentieth century. This recent provenance may seem to support the accusation that contemporary materialism owes its popularity more to fashion than to any serious argument. 'If the case is so substantial', anti-materialists can ask, 'how come it took so long for philosophers to appreciate it?' I take this to be a good historical question. But I think there is also a good historical answer: namely, that a key premiss in the argument for materialism rests on empirical evidence that only became clear-cut during the course of the twentieth century.

However, I shall not complicate the analysis of this chapter by overlaying it with historical commentary. The issues are complicated enough without the added burden of tracing historical strands. Accordingly, this chapter will focus on the structure of the argument for materialism, not its history. For those who are interested in the historical dimension, the Appendix at the end of this book discusses the history of the causal argument, and in particular the question of why it has become persuasive only recently.

1.2 *The Causal Argument*

Let me now outline what I take to be the canonical argument for materialism. Setting to one side all complications, which can be discussed later, it can be put as follows.

> Many effects that we attribute to conscious causes have full physical causes. But it would be absurd to suppose that these effects are caused twice over. So the conscious causes must be identical to some part of those physical causes.

To appreciate the force of this argument, consider some bodily behaviour which we would standardly attribute to conscious causes. For example, I walk to the fridge to get a beer, because I consciously feel thirsty. Now combine this example with the thought that, according to modern physical science, such bodily movements are fully caused by prior physical processes in brains and nerves. The obvious conclusion is that the conscious thirst must be identical with some part of those physical processes.

Let me now lay out the above argument more formally. This will help us to appreciate both its strengths and its weaknesses.

As a first premiss, take:

(1) Conscious mental occurrences have physical effects.

As I said, the most obvious examples are cases where our conscious feelings and other mental states cause our behaviour.

Now add in this premiss ('the completeness of physics' henceforth):

(2) All physical effects are fully caused by purely *physical* prior histories.[2]

[2] What about quantum indeterminacy? A stricter version of (2) would say that the *chances* of physical effects are always fully fixed by their prior physical histories, and would reformulate the rest of the argument accordingly (with (1) then as 'Conscious mental occurrences affect the *chances* of physical effects', and so on). I shall skip this complication in most of what follows.

In particular, this covers the behavioural effects of conscious causes to which our attention is drawn by premiss 1. The thought behind premiss 2 is that such physical behaviour will always be fully caused by physical contractions in your muscles, in turn caused by electrical messages travelling down your nerves, themselves due to physical activity in your motor cortex, in turn caused by physical activity in your sensory cortex, and so on.

At first sight, premisses 1 and 2 seem to suggest that a certain range of physical effects (physical behaviour) will have two distinct causes: one involving a conscious state (your thirst, say), and the other consisting of purely physical states (neuronal firings, say).

Now, some events are indeed overdetermined in this way, like the death of a man who is simultaneously shot and struck by lightning. But this seems the wrong model for mental causation. After all, overdetermination implies that even if one cause had been absent, the result would still have occurred because of the other cause (the man would still have died even if he hadn't been shot, or, alternatively, even if he hadn't been struck by lightning). But it seems wrong to say that I would still have walked to the fridge even if I hadn't felt thirsty (because my neurons were firing), or, alternatively, that I would still have gone to the fridge even if my neurons hadn't been firing (because I felt thirsty). So let us add the further premiss:

(3) The physical effects of conscious causes aren't always overdetermined by distinct causes.

Materialism now follows. Premisses 1 and 2 tell us that certain effects have a conscious cause and a physical cause. Premiss 3 tells us that they don't have two distinct causes. The only possibility left is that the conscious occurrences mentioned in (1) must be identical with some part of the physical causes mentioned in (2). This respects both (1) and (2), yet avoids the implication of overdetermination, since (1) and (2) no longer imply *distinct* causes.

1.3 *The Ontology of Causes*

The causal argument focuses on the way in which conscious occurrences operate as *causes*. It says that conscious *causes* must be identical to physical *causes*. However, there are different philosophical theories of causation, and in particular about the kinds of things that can feature as causes. On one view, causes are facts, or instantiations of properties. Candidate causes on this view would be *my being in pain*, or *my having active nociceptive-specific neurons*. On an opposed view, causes are basic particulars, or *events*, abstracted from any conscious or physical properties they might have. The causal argument as stated above will generate different conclusions, depending on which view of causation you adopt. In particular, it will generate a stronger conclusion on the former view, that causes are facts, than on the latter view, which has causes as basic particulars. Still, this will be of no great moment, since a rephrasing of the argument will still allow us to generate the stronger conclusion, even on the assumption that causes are basic particulars.

Let me take this a bit more slowly. I myself favour the view that causes are facts (cf. Mellor 1995). A restricted variant of this view, which will perhaps be more familar to some readers, is that causes are instantiations of properties by particulars, or 'Kim-events' (cf. Kim 1973). In what follows, I shall standardly use the term 'state' to refer to this kind of item—that is, to the possession of a property by some particular. Now, on the view that causes are facts (Kim-events, states) the causal argument given above implies that conscious *properties* (being thirsty, say) must be identical with physical *properties* (having a certain brain feature). For, in requiring that conscious causes be identical with physical causes, the argument will now require that conscious facts (Kim-events, states)—such as that *I am thirsty*, say—are identical to certain physical facts (Kim-events, states)—*I have a certain brain feature*, say—and these two facts (Kim-events, states) cannot be identical unless the properties they involve—being thirsty, having that brain feature—are themselves identical.

The alternative view of causation is that causes are basic particulars (cf. Davidson 1980). Then the causal argument, as phrased above, won't itself carry you to the identity of conscious and physical properties, since the identity of conscious Davidson-events with physical ones requires only the far weaker conclusion that the relevant conscious and physical properties are instantiated in the same particular, not that the properties themselves are identical.

Still, as I said, we can rephrase the argument so as to regenerate the stronger conclusion. Let us take premiss 1′ to be the claim that all conscious events cause some physical events *in virtue of* their conscious properties; premiss 2′ says that all physical events are caused by prior physical events *in virtue of* the latter's physical properties; and premiss 3′ says that the physical effects of conscious causes aren't always caused twice over, *in virtue of* two different properties of the prior circumstances. In order to make these consistent, we then need once more to identify the conscious properties of the causes with their physical properties.

The causal argument as presented in the last section thus argues for the identification of conscious properties with physical properties. It is worth noting at this stage, however, that this argument for property identity proceeds on an abstract, existential level, and is not concerned with any detailed identifications. It tells us that each conscious property must be identical with *some* physical property, but it doesn't tell us *which* specific physical property any given conscious property may be identical with.[3]

To establish any such specific property identity, more detailed empirical information is needed. It is not enough to know that conscious causes can always be identified with *some* part of the full physical histories behind their effects. To pin down specific property identities, we need more detailed evidence about correlations between specific conscious properties and the different parts of those physical histories. We need to know that pain, say, or thirst, or seeing an elephant, are found when such-and-such brain areas are active, but not when others are. In Chapter 7 I shall consider this kind of detailed research, and the kinds of results it can be expected to bring. But for the moment, I shall concentrate on the more abstract existential claim that every conscious property must be identical with *some*, as-yet-to-be-identified physical property.

[3] Similarly, on the alternative construal of the causal argument to be developed in section 1.7, we will have an abstract argument for the identity of conscious properties with material properties (even if not with strictly physical properties), but again this argument on its own will not tell us which material properties any specific conscious property should be identified with.

We can usefully think of this abstract claim and the detailed correlational research as complementing each other. The abstract claim doesn't by itself tell us which physical property a given conscious property should be paired up with. And the correlational research, while promising to establish specific pairings, can't by itself establish that the paired properties are *identical*, as opposed to regularly accompanying each other. The abstract claim is important, then, since it is needed to license the move from detailed empirical correlations to property identifications. It tells the empirical researchers that conscious properties aren't just *correlated* with the physical properties they are regularly found with, but must be identical with them.

1.4 *Epiphenomenalism and Pre-established Harmony*

All this assumes, however, that the abstract claim does follow from the causal argument. Let us now examine this argument more closely.

As laid out above, the causal argument seems valid.[4] So, to deny the conclusion, we need to deny one of the premisses. All of them can be denied without contradiction. Indeed, all of them have been denied by contemporary philosophers, as we shall see. At the same time, they are all highly plausible, and their denials have various unattractive consequences.

Let me start with premiss 1. This claims that, as a matter of empirical fact, particular conscious states have particular physical effects. This certainly seems plausible. Doesn't my conscious thirst cause me to walk to the fridge? Or, again, when I have a conscious headache, doesn't this cause me to ingest an aspirin?

Still, the possibility of denying this premiss is familiar enough, under the guise of 'epiphenomenalism' or 'pre-established harmony'.

The first philosopher to embrace this option was Leibniz. Unlike

[4] However Sturgeon (1998) argues that the argument trades on an equivocation between the everyday sense of 'physical' (in premiss 1) and a quantum-theoretical sense (in premiss 2). I shall comment on Sturgeon's claim in section 1.10 below.

most other philosophers prior to the twentieth century, Leibniz was committed to the causal completeness of physics (see Appendix). But he was not prepared to accept the identity of mind with brain. So he opted for a denial of our premiss 1, and concluded that mind and matter cannot really influence each other, and that the appearance of interaction must be due to *pre-established harmony*. By this Leibniz meant that God must have arranged things to make sure that mind and matter always keep in step. In reality, they do not interact, but are like two trains running on separate tracks. But God fixed their starting times and speeds so as to ensure they would always run smoothly alongside each other.

Some contemporary philosophers (for example, Jackson 1982) follow Leibniz in avoiding mind-brain identity by denying premiss 1. But they prefer a rather simpler way of keeping mind and matter in step. They allow causal influences 'upwards' from brain to mind, while denying any 'downwards' causation from mind to brain. This position is known as *epiphenomenalism*. It respects the causal completeness of physics, in that nothing non-physical causally influences the physical brain. But it avoids the theological complications of Leibniz's pre-established harmony, by allowing the brain itself to cause conscious effects.

Epiphenomenalism is not a particularly attractive position. For a start, it would require us to deny many apparently obvious truths, such as that my conscious thirst caused me to fetch a beer, or that my conscious headache caused me to swallow an aspirin. According to epiphenomenalism, my behaviour in both these cases is caused solely at the physical level. These physical causes may be accompanied by conscious thirst or a conscious headache, but these conscious states no more cause resulting behaviour than falling barometers cause rain.[5]

[5] Chalmers (1996: esp. 134–6), following Russell (1927) and Lockwood(1989), argues that there is a way for dualism to avoid this epiphenomenalist inefficacy while respecting the completeness of physics. This is to identify phenomenal properties with the *intrinsic* properties of the physical realm. Chalmers's idea is that physical science picks out properties like mass and charge only extrinsically, via their relations to observable features of the world. So maybe phenomenal properties can be identified with the intrinsic nature of such properties, suggests Chalmers, and thereby have their causal efficacy restored. This seems an entirely

That epiphenomenalism has these odd consequences is not in itself decisive. The theoretical truth can often overturn claims which were previously regarded as the merest common sense. Moreover, there is nothing incoherent about epiphenomenalism. As I shall have occasion to stress in what follows, there is nothing conceptually contradictory in the idea of conscious states which exert no causal powers themselves. Still, epiphenomenalism is surely an empirically implausible position, by comparison with the materialist view that conscious states are simply identical to brain states.

If epiphenomenalism were true, then the relation between mind and brain would be like nothing else in nature. After all, science recognizes no other examples of 'causal danglers', ontologically independent states with causes but no effects. So, given the choice between epiphenomenalism and materialism, standard principles of scientific theory choice would seem to favour materialism. If both views can accommodate the empirical data equally well, then ordinary scientific methodology will advise us to adopt the simple view that unifies mind and brain, rather than the ontologically more profligate story which has the conscious states dangling impotently from the brain states.

There remains the possibility that the anti-materialist arguments to be examined later will show that conscious mind and brain *cannot* be identical. If this is so, then one of the premisses of the causal argument must be false. And in that case premiss 1 seems as likely a candidate as any. Certainly most contemporary philosophers who are persuaded by the anti-materialist arguments have opted for epiphenomenalism and the denial of premiss 1, rather than for any other way out of the causal argument.

sensible view to me. But, *pace* Chalmers, I would say that it is simply a version of materialism. My reaction is that the intrinsic features of the physical world with which Chalmers wants to identify phenomenal properties are themselves simply basic physical properties. Thus I am happy to agree with Chalmers that scientific theory picks out these intrinsic physical properties only via descriptions which refer to observable features of the world. Moreover, I agree that conscious properties should be identified with arrangements of such intrinsic physical properties, and thus that it is like something to have these arrangements of intrinsic properties. Indeed, I find it hard to see what a sensible materialism could amount to, except this combination of views. So, from my point of view, Chalmers's suggested position is simply the optimal formulation of materialism.

But this does not invalidate the criticisms I have levelled against epiphenomenalism. My concern at the moment is not to prejudge the anti-materialist case, but merely to assess the causal argument. And the point remains that, in the absence of further considerations, it seems clearly preferable to identify mind with brain than to condemn conscious states to the status of causal danglers. It may be that further anti-materialist considerations will yet require us to reconsider this verdict, but so far we have seen no reason to deny premiss 1, and good reason to uphold it.

Before leaving the issue of epiphenomenalism, it may be worth addressing some more local worries about premiss 1. Even if the blanket epiphenomenalist refusal to credit *any* conscious states with physical effects is methodologically unattractive, there may be some more specific reasons for doubting whether particular sorts of conscious states have the physical effects they are normally credited with. In particular, I am thinking here of conscious *decisions*, and doubts about their causal efficacy arising from the experimental results associated with Benjamin Libet, and of conscious states which are *representational*, and doubts about their causal efficacy arising from the possibility that they may have 'broad contents'. Let me deal with these in turn.

In a series of well-known experiments, Libet asked subjects to decide spontaneously to move their fingers, and simultaneously to note the precise moment of their decision, as measured by a large stop-watch on the wall. Libet also used scalp electrodes to detect the onset of motor cortical activity initiating the finger movement. Amazingly, he found that this neural activity started a full $\frac{1}{3}$ to $\frac{1}{2}$ second *before* the subjects were aware of making any conscious decision (Libet 1993).

At first sight, this certainly suggests that such conscious decisions are epiphenomenal with respect to the actions we normally attribute to them: since the conscious decisions come later, it looks as if they must be effects of, rather than identical with, the brain processes that give rise to the action. But in fact this interpretation is not clear-cut. Libet himself points out that the conscious decisions still have the power to 'endorse' or 'cancel', so to speak, the processes initiated by the earlier cortical activity: no action will result if the action's execution is consciously countermanded. Given this, it seems that the conscious decision is part of the cause of the finger movement after all. The initial cortical activity does not determine the finger movement on its own, but only puts the motor cortex in a state of 'readiness', which leads to action in just those cases where the conscious decision is added. This then allows us to reason, as before, via the causal argument, that conscious decisions could not play a part in so influencing physical movements, were they not themselves physical.

In any case, even if conscious decisions did *not* contribute causally to the actions normally attributed to them, it would not follow that they had no physical effects of *any* kind. For instance, they will still presumably be causes of the sounds I make, or the marks I put on paper, when I later *report* my earlier conscious decisions. So they will still satisfy premiss 1, which requires only that conscious causes have *some* physical effects, and not that they have all the physical effects with which they are normally credited by common sense. So once more the causal argument will run.

The other worry concerned the possibility of conscious states with 'broad' representational contents. The possession of such 'broad contents' hinges on matters outside subjects' heads. For example, Hilary Putnam suggests that the representational state *thinking about water* hinges on what natural kind is actually water in your environment, and Tyler Burge argues that *thinking about arthritis* hinges on facts about other members of your community (Putnam 1975, Burge 1979, 1982).

Now the worry, in the present context, is that if any conscious states are representational in this broad way, then this will not sit happily with premiss 1's claims about causal efficacy. For how can states which hinge on matters outside your head exert a causal influence on your bodily movements? Surely your bodily movements are causally influenced solely by matters inside your skin, not by how matters are outside you.

The possibility of broadly representational conscious states raises any number of tricky issues, not all of which I can pursue here (though see section 7.7 below). However, they seem to me to pose no real threat to the causal argument for materialism. Let me content myself with two comments.

First, I am open to the possibility that some, indeed all, conscious states may be essentially representational (cf. n. 1 above); moreover, it seems plausible that representation in general is a broad matter. Even so, it would seem odd to allow that conscious properties in particular, as opposed to representational properties in general, can depend on broad matters outside the skin. Could two people really be internally physically identical, yet nevertheless *feel* different, because things are different outside them? (Cf. Introduction, n. 2.) Given this, the natural strategy for those who seek to equate some (or all) conscious properties with representational properties is to shear off some species of narrow representation from the general run of broad representational properties, and to equate representational conscious properties with these narrow representational properties. And then, to return to the matter at hand, there will cease to be any reason to doubt that these conscious properties have physical effects such as bodily movements, however it may be with representational properties in general.

Second, even if you do wish to insist that some conscious properties are indeed broadly representational (a possibility to which I shall return in section 7.7), it will not follow that such broad conscious properties do not cause *any* physical effects. For they may have physical effects *outside* my body. For example, my consciously thirsting for water might affect *which liquid* I put into a glass, and my consciously worrying about arthritis might affect *where the doctor will poke me* when I complain of it. If this is right, then the causal argument will run as before, and imply that any such broad conscious properties must also be identical with physical properties, if their instantiations are to have such physical effects—though these physical properties will now presumably stretch outside bodies, as well as inside.

1.5 *Accepting Overdetermination*

There remain the two other premisses to the causal argument. It will be convenient to relegate the discussion of premiss 2, the completeness of physics, to the last section of this chapter and the

Appendix. So let me now briefly consider premiss 3, the one ruling out overdetermination.

To reject this premiss is to accept that the physical effects of mental causes are always overdetermined by distinct causes. This is sometimes called the 'belt and braces' view (make doubly sure you get the effects you want), and is defended by D. H. Mellor (1995: 103–5).

At first sight, this position seems to have the odd consequence that you would still have gone to the fridge for a beer even if you hadn't been thirsty (because your cortical neurons would still have been firing), and that you would still have gone to the fridge even if your cortex hadn't been firing (because you would still have been thirsty). These counterfactual implications seem clearly mistaken.

However, defenders of the belt and braces view maintain that such implications can be avoided. They argue that the distinct mental and physical causes may themselves be strongly counterfactually dependent (that is, they hold that, if you hadn't been thirsty, your sensory neurons wouldn't have fired either, and vice versa).

Still, this then raises the question of *why* such causes should always be so counterfactually dependent, if they are ontologically distinct.[6] Why wouldn't my neurons have fired, even in the absence of my conscious thirst? Similarly, why shouldn't I still have been thirsty, even if my neurons hadn't fired? Now, it is not impossible to imagine mechanisms which would ensure such counterfactual dependence between distinct causes. Perhaps the conscious thirst occurs first, and then invariably causes the cortical activity, with both causes thus available to overdetermine the behaviour. Alternatively, the cortical activity could invariably cause the thirst. Or, again, the conscious decision and the cortical activity might be joint effects of some prior common physical cause. But such mechanisms, though conceptually coherent, seem highly implausible, especially given that they need to

[6] Note that this is only a problem if the causes are genuinely ontologically distinct, and not if they are merely related as role state and physical realizer. As we shall see in the next section, the existence of 'two' causes in this latter sense does not threaten overdetermination, precisely because of their ontological interdependence. So I have no objection to versions of the belt and braces view which intend only parallel causes in this weak sense. Cf. Segal and Sober 1991.

ensure that the conscious state and the brain state *always* accompany each other.

The relevant point is analogous to one made in the last section. We don't find any 'belt and braces' mechanisms elsewhere in nature—that is, mechanisms which ensure that certain classes of effects invariably have two distinct causes, each of which would suffice by itself. As with the epiphenomenalist model, a belt and braces model requiring such peculiar brain mechanisms would seem to be ruled out by general principles of scientific theory choice. If the simple picture of mental causation offered by materialism accommodates the empirical data as well as the complex mechanisms required by the belt and braces option, then normal methodological principles would seem to weigh heavily against the belt and braces view.

As with the corresponding argument for epiphenomenalism, this appeal to principles of scientific theory choice is defeasible. Perhaps in the end the anti-materialist arguments will force us to accept mind–brain distinctness. In that case, the belt and braces view might be worth another look. True, it is even more Heath-Robinsonish than epiphenomenalism. On the other hand, it does at least have the virtue of retaining the common-sense view that conscious states characteristically cause behaviour. In any case, my present purpose is not to decide this issue finally, but only to point out that, as things stand so far, we have good reason to uphold premiss 3, and none to deny it.

1.6 *Functionalism and Epiphobia*

Many contemporary philosophers will feel that the causal argument as elaborated so far is rather too strong. This argument has claimed that conscious properties are identical to *physical* properties. But the majority of contemporary materialists would probably prefer to identify mental properties in general, and conscious properties in particular, with physically realized *functional* properties, or properties which *supervene* on physical properties, or perhaps properties which are *disjunctions* of physical properties, rather than with strictly physical properties themselves.

Let me start with functional properties. I shall come back to the other possibilities in a moment. A functional property is a higher-order property-of-having-some-property-which-satisfies-condition-R, where R specifies some requirement on an instantiation of a first-order property. In line with this, 'functionalism' in the philosophy of mind is the view that any given mental property should be identified with some property-of-having-a-first-order-property-which-bears-certain-causal-relationships-to perceptual inputs, behavioural outputs, and other mental states. For example, the property of being in pain might be identified, at first pass, with the property-of-having-some-property-which-arises-from-bodily-damage-and-gives-rise-to-a-desire-to-avoid-the-source-of-that-damage.

The advantage of this functionalist account of mental states is that it allows beings who have quite different instrinsic physical properties nevertheless to share mental properties. For example, it seems plausible that octopuses, whose neurology is physically quite different from human neurology, can nevertheless share the property of being in pain with humans. But, if this is so, the property of being in pain cannot be identical with any physical property, for no suitable physical property will be common to humans and octopuses. On the other hand, both humans and octopuses will share the higher-order property-of-having-some-property-which-arises-from-bodily-damage-and-gives-rise-to-a-desire-to-avoid-the-source-of-that-damage. The physical properties which play this role will be different in the two cases, but the higher-order property itself will be common to octopuses and humans.

Now, how does functionalism stand with respect to the causal argument? If we take the causal argument at face value, then they seem inconsistent. For, as we have seen, the causal argument promises to establish that conscious properties are identical with strictly physical properties, which is just the claim that functionalism is designed to avoid.

This tension with the causal argument puts functionalism under some pressure. If functionalism is inconsistent with the causal argument, it must deny one of its premisses. And, on reflection, it could well be held to deny premiss 1, the one that says that conscious causes have physical effects. For, if conscious properties are not identical with physical properties, but rather with certain higher-order properties, then conscious causes will not be identical with the physical causes which premisses 2 and 3 tell us are the only causes of behavioural effects. So it would seem to follow that conscious states don't cause behavioural effects after all. This line of thought is sometimes said to generate 'epiphobia', a condition in which functionalists are overcome with anxiety about how their view differs from epiphenomenalism.

For this reason, and perhaps others, some philosophers have recently become uneasy about 'higher-order' properties. They object that it is profligate to posit substantial new properties for every way of *characterizing* objects as possessors of some (first-order) property which R (cf. Kim 1998: ch. 4).

I have some sympathy with this point of view. However, it is important to realize that, even if we reject higher-order properties on these grounds, the underlying dilemma highlighted by functionalism remains. For we will still need to decide whether conscious properties should be identified (a) with those strictly physical properties whose instantiations are paradigm physical causes, yet are not shared by humans and octopuses, or (b) with other first-order properties of a kind which can be shared by humans and octopuses, but are in danger of being outcompeted as serious causes.

Suppose, to illustrate the point, that we admit no properties except genuinely first-order properties. But suppose that we also continue to feel the pull of the thought that both humans and octopuses can be in pain. Given the physical differences between humans and octopuses, we might seek to respect this thought by construing *pain* as a *disjunctive* condition, requiring P_1 *or* P_2 *or* . . . where the various P_is are the different strictly physical properties which are causally active when different beings are in pain. But now epiphobia returns to trouble us once more. For my human arm movement is presumably caused by my human P_1 (my nociceptive-specific neurons firing, say). But P_1 itself isn't identical with the disjunction P_1 *or* P_2 *or* . . .—that is, with pain. So, if P_1 causes my movement, the disjunction presumably doesn't, and thus it seems to follow once more that the property of being in pain is inefficacious. The dilemma remains: if you want to have different creatures sharing pain, then you seem to end up rendering pains causally inefficacious.

A similar point can be made about views which replace higher-order functional properties, not by disjunctions of physical properties, but by properties which 'supervene' on physical properties. (For readers unfamiliar with this notion, it is explained in section 1.8.) This alternative will again leave us with the choice between identifying conscious properties with (a) physical properties themselves, or (b) with the properties which supervene on physical properties. And again we will face the dilemma that only (b) seems to allow physically different beings to share conscious properties, but only (a) seems to allow conscious properties to be causally efficacious.

In the next section I shall consider whether this dilemma can be resolved. However, it would be tiresome to have to address the issue separately for all the different ways in which conscious properties can be identified with properties which are not strictly physical—that is, for functional higher-order properties *or* disjunctions of physical properties *or* supervenient properties. So let me adopt the general term 'higher' property to cover all these alternatives. Correspondingly, when I speak of a 'higher' property being 'realized' by a physical property, I shall mean either that a functional higher-order property is instantiated because some physical property is, or that a disjunction of physical properties is instantiated because one of its disjuncts is, or that a supervenient property is instantiated because some physical property which determines it is.[7]

[7] It might seem that for completeness I should also consider the possibility that conscious properties can be identified with macro-properties that are *composed of* physical micro-properties. However, the mereological notion of a macro-whole being composed of micro-parts seems to me orthogonal to the notion of one kind of property being *realized* by another kind of property. I would say that composition can occur *within* the strictly physical, and accordingly that a macro-property composed of strictly physical micro-properties is itself a strictly physical property. So identifying conscious properties with properties composed of strictly physical properties is itself to identify them with strictly physical properties, not with some species of 'higher' property realized by physical properties. There are interesting questions about the compositional relation between micro-parts and macro-wholes—in particular, do the parts have some kind of causal primacy over wholes?—but they are independent of the arguments in this book. (For convincing arguments against the causal primacy of micro-parts, see Hüttemann, forthcoming.)

1.7 *A Possible Cure for Epiphobia*

Perhaps there is a cure for epiphobia. We don't have to agree that the only respectable kind of *causation* involves strictly physical causes having physical effects. For it is arguable that there is a perfectly normal sense of 'cause' in which higher states cause the effects that their realizers cause. On this account, even if pain is a higher property, differently realized in octopuses and humans, my taking an aspirin can still be caused by the pain in my head, in virtue of being caused by whichever strictly physical state realizes that pain in me.

If we adopt this generous notion of causation, functionalism becomes consistent with premiss 1 of the causal argument after all. The fact that mental states are not identical with strictly physical states does not mean that they cannot cause the behaviour which is caused by those strictly physical states. In the generous sense of 'cause', they will do so as long as they are higher states which are realized by those strictly physical states.

Indeed, if we look at things in this way, we in effect have another version of the causal argument, one which reads 'cause' generously throughout, and which ends up with the conclusion that conscious properties, if not strictly physical properties, must at least be *physically* realized higher properties. The argument now runs:

(1*) Conscious causes have physical effects, at least in the generous sense.

(2) All physical effects are fully caused by purely physical prior histories.

(3) The physical effects of conscious causes aren't overdetermined by distinct causes.

And the conclusion is now that:

(4*) Conscious causes must at least be higher states which are realized by the *physical* causes of their physical effects.

For otherwise (3) would be violated, with the physical effects of conscious causes being caused twice over, first by their conscious causes as in (1′), and second by the distinct physical causes guaranteed by (2).

Note here how we do *in a sense* end up with two causes of the relevant behavioural effects. For we now have both (a) the higher state with which we are now identifying the conscious state and (b) the realizing physical state which directly causes the behavioural result. (Cf. Segal and Sober 1991.)

But the important point is that these two 'causes' are not now ontologically *distinct*, and so do not genuinely overdetermine any resulting behaviour. The higher cause is present only in virtue of the physical cause which realizes it. In the circumstances, the one would be absent if the other were. And because of this, we have no trouble with the counterfactuals which would be indicative of genuine overdetermination. It is *not* true that the behavioural result would still have been caused even if the physical realizer had been absent, for the higher state would then have been absent too;[8] and similarly, if the higher state had been absent in some particular case, there would again have been no alternative cause for the behavioural result, since the physical realizer would have had to be absent too.

Note that it is not essential to this rejigged version of the causal argument that we *start* with any assumption that conscious states are higher states. I shall be considering alternative arguments for materialism shortly, and in particular a form of argument that begins with a functionalist assumption of just this sort, taken to be derivable a priori from the structure of our concepts of conscious states. If you begin with this kind of a priori functionalism, a variant of the causal argument can still serve an important purpose: namely, that of establishing that higher mental states are *physically* realized, as opposed to being realized by some distinctive non-physical

[8] Mightn't this be false? Since different physical properties can realize a given higher state, isn't it possible that a different realizer would have been present if the actual realizer had been absent? There are some delicate issues here (cf. Yablo 1992). But for present purposes the significant point is that a given creature certainly wouldn't have had a given higher property if it hadn't had the kind of physical property that features as a realizer when it itself has that higher property. For example, I certainly wouldn't have been in pain if I didn't have the kind of physical property that realizes pain in me (since, after all, there isn't any question of my having the realizer that occurs in octopuses or other creatures).

mind-stuff (cf. Lewis 1966). But this is not how I am thinking of the rejigged causal argument.

Rather, I intend it to establish *both* that conscious states must at least be higher states, if not strictly physical, *and* that they must be physically realized. That is, I am taking the identity of conscious properties with higher (or physical) properties to be the *conclusion* of my argument, not a premiss. The premisses are simply (1*), (2), and (3), which make no claims, a priori or otherwise, about the specific nature of conscious states, and the conclusion is that, if conscious properties are not strictly identical with physical properties, then they must at least be identical with higher-properties-which-are-physically-realized, otherwise we will be driven to deny that conscious states cause their effects in any sense or, alternatively, to accept that those effects are genuinely overdetermined by quite distinct causes.[9]

So far in this section I have shown how functionalism and other 'higher property' versions of materialism can respect the premisses of the causal argument, and indeed can use the rejigged version as an argument in their favour. However, I have not intended this as a defence of such views. This is because I am not sure whether they can really be cleared of the charge of epiphenomenalism.

The issue here hinges on whether we can seriously allow that higher states *cause* what their realizers cause. I am not sure what to say about this. Sometimes I think that this is not a serious notion of causation, and certainly not one which does justice to the way in which my thirst causes me to drink a beer. Surely, one feels, my thirst itself is efficacious in getting me to move, in just the same strict way as physical causes produce their effects, and not merely in the second-hand sense that it is realized by some other state which causes in this strict sense.

When I am in this mood, I am inclined to read the causal argument as employing a strict notion of causation throughout, and in particular in premiss 1's assertion that conscious states *cause* physical

[9] I take this construal of the causal argument to rebut Tim Crane's (1995) complaint that the causal argument must assume a form of mental causation which will then be denied by any functionalist-style higher-property version of materialism.

effects. This then drives me to the conclusion that conscious properties must be identical to strictly physical properties, and that any higher properties are merely epiphenomenal. The cost of this strictly physicalist position, of course, is that I will not share conscious properties with octopuses or other physically distinct beings. But perhaps this isn't as bad as it seems. After all, it doesn't mean that octopuses don't have any conscious properties at all. And I will still share *some* properties with them, albeit not the conscious properties that strictly cause our respective behaviours. (In particular, I will share some higher properties, which is perhaps why we can both count as in 'pain'.)

At other times I feel less fussy about causation. In particular, I sometimes worry that we will be left with precious few causes, if we are going to hold that higher states are pre-empted as causes whenever they have realizers in virtue of which they cause. For, if applied strictly, this principle threatens to block the causal efficacy of even such eminently respectable causal states as pressures and temperatures. After all, on any particular occasion the effects of temperatures and pressures will also be caused by specific molecular movements. These specific movements will *realize* the relevant pressures or temperatures, but won't be *identical* to them, since the pressures and temperatures can also be realized differently. So the pressures and temperatures won't count as causes, if they can't cause what their realizers cause.

This seems odd, and argues against dismissing higher states from the realm of serious causes, and in favour of a generous reading of premiss 1. On this reading, my thirst will still be a serious cause of my going to the fridge, even if it has a realizer in virtue of which it causes. And then the causal argument will simply yield the conclusion that it must be a physically realized higher state, not that it must be strictly physical itself.

As I said, I am not sure what to say about this issue. It is a complicated matter, and it is not clear how best to resolve it. Fortunately, nearly all the arguments in the rest of this book will be insensitive to this issue. We can identify conscious properties either with strictly physical properties or with physically realized higher properties. Whichever choice we make, we will still have an identity between conscious properties and properties which are innocent of any of the obscurities which surround consciousness. This is the important point, and beyond that it will not matter too much whether conscious properties are identified with strictly physical or with physically realized higher properties.[10]

1.8 *Intuition and Supervenience*

Let me now distinguish the causal argument we have been examining from some other ways of defending materialism that can be found in the recent literature.

To start with, it is sometimes suggested that materialism about consciousness can be established by a priori intuition alone. This is a feeble thought, as will become clear shortly, but its deficiencies have sometimes been obscured by the fashion for thinking of materialism about the mental in terms of 'supervenience': that is, in terms of the doctrine that any two beings who share all physical properties must also share all mental properties.

I myself find the notion of supervenience more trouble than it's worth. The notion of supervenience has proved far less straightforward than it at first seemed, and has generated a huge amount of technical literature (mostly focusing on the 'must' in 'if . . . physically identical . . . *must* also be mentally identical'). I would argue that any benefits offered by the notion of supervenience are more easily gained simply by identifying mental properties directly with higher-order properties or disjunctions of physical properties. Accordingly, the notion of supervenience will not play a prominent part in the rest of this book.

I mention it here only because supervenience formulations of

[10] In Chapter 7 I shall return to the choice between functionalist-style higher-property views and strict physicalism, and consider whether empirical research into conscious properties can help resolve the issue. By that stage I shall also have developed an extensive account of the structure of our *concepts* of conscious properties. But I shall argue that none of this delivers a resolution, and my eventual conclusion will be that our concepts of conscious properties are vague, in that it is indeterminate whether they refer to higher properties or to the physical properties which realize them in humans.

materialism can create the spurious impression that materialism is a purely intuitive matter. After all, there is a sense in which a priori intuition does tell us that the conscious realm supervenes on the physical realm. Everybody has strong intuitions about the correlation between mind and brain. If I made a molecule-for-molecule physical copy of you using a Star Trek-style teletransporter, for example, wouldn't your physical twin automatically have all the same feelings that you have?

However, this intuition-based supervenience falls far short of anything worth calling materialism. To see why, note that the teletransporter thought-experiment is consistent with epiphenomenalism: perhaps the copy feels like the original simply because its brain states causally generate extra conscious states, in just the same way as the original's brain states do. Here the conscious states would be distinct from the brain states, but would regularly accompany them, in virtue of laws by which brain states cause conscious states. A merely epiphenomenalist mind–brain correlation like this clearly doesn't amount to materialism.

In the technical terminology into which we are forced by the apparatus of supervenience, the point is that the teletransporter thought-experiment shows only that physical identity guarantees conscious identity across *natural* possibilities, possibilities which share all our natural laws, including any brain–mind epiphenomenal laws. However, to establish a supervenience amounting to genuine materialism, we would need to show that physical duplicates couldn't *possibly* be mentally different, whatever the laws of nature, not just that they aren't different in worlds which do share our laws. We need to establish supervenience of the mental across all metaphysically possible worlds. Only this promises to ensure that the mental is ontologically inseparable from the physical, and not just correlated with it.

If you find this obscure, the point can be put more directly in terms of ontological relations between mental and physical properties. Mere supervenience across naturally possible worlds doesn't amount to materialism, because it doesn't rule out the epiphenomenalist possibility that conscious properties are ontologically quite distinct from physical properties, albeit constantly correlated with them by epiphenomenal laws in this actual world and those nearby worlds that share our natural laws. Supervenience across all possible worlds, on the other hand, does arguably suffice for materialism, precisely because an ontological dependence of mental on physical properties seems the only thing that will enable physical identity to *necessitate* mental identity, whatever laws may obtain.

Now that we see which version of supervenience is required to ensure genuine materialism, it should be clear that intuition alone will fail to deliver the materialist goods. It is not at all intuitively obvious that physical duplicates must *necessarily* be conscious duplicates, that a physical doppelganger couldn't *possibly* have different experiences. Even a dyed-in-the-wool materialist, like myself, feels the pull of the intuition that there could be a 'zombie', say, who is physically just like me but has no feelings—in a possible world, so to speak, where any epiphenomenal laws relating brain states to conscious states have broken down.

It may in fact be true that zombies are impossible, and indeed this is something for which I shall argue at length in due course. My present point is only that a priori intuition alone cannot establish their impossibility. If anything, it suggests just the opposite.

1.9 *An Argument from A Priori Causal Roles*

Let me now consider one further form of argument for materialism. This shares some of the structure of the causal argument. But in place of premiss 1 or 1*, which simply states that, as a matter of fact, conscious causes have physical effects, this argument appeals instead to a putative a priori analysis of our *concepts* of conscious states.

According to this line of thought, our concepts of conscious states, like *pain*, or *thirst*, are each associated a priori with the specification of some causal role linking that state to physical causes and effects (cf. Lewis 1966). So, as above, our concept of pain would be linked a priori with bodily damage as cause and a desire to avoid the source of the pain as effect. Again, our concept of thirst would be linked to lack of water as cause and a desire to drink as effect.

This kind of a priori analysis can then be plugged into the rest of the causal argument, so to speak, to deliver the materialist conclusion. The a priori analysis tells us that conscious states have a causal role, and hence have physical effects. The completeness of physics tells us that these physical effects must have full physical histories. The denial of overdetermination tells us that these physical effects aren't caused twice over. Thus, once more we reach the conclusion that conscious states cannot be ontologically distinct from the physical causes of their physical effects.[11]

There may seem no great distance between the causal argument discussed earlier and this argument appealing to a priori analyses of our concepts of conscious states. However, it is crucially important that the causal argument discussed earlier rests on no such a priori assumptions. While that causal argument assumed that conscious causes have physical effects, it offered this as a straightforward empirical truth, not as a conceptual matter.

In line with this, note how I have been happy to allow the *conceptual* possibility that conscious states may lack effects altogether. This point arose earlier in my discussion of epiphenomenalism. My reason for dismissing epiphenomenalism was not that its denial of mental efficacy violated any conceptual truths, but simply that it amounted to an empirically far less plausible story than the simple identities postulated by materialism.

I shall have a lot more to say about our concepts of conscious states in what follows. Without wanting to pre-empt that analysis, let me simply say at this stage that it will amply confirm that there are no

[11] Interestingly, this form of argument can be used to deliver either a functionalist version of materialism—conscious states are identical with physically realized higher-order states—or a more strictly physicalist conclusion—conscious states are identical with the physical states themselves. We will get the former conclusion if we take it to be given a priori that conscious properties are higher-order properties—that is, that they are identical to properties like the-property-of-having-some-property-which-plays-such-and-such-a-causal-role. The overall argument will then lead us to the conclusion that these higher-order properties must be realized *physically*, as opposed to being realized by some special mind-stuff. But we can also take the causal roles associated a priori with concepts of conscious states to fix reference to whichever first-order properties actually play those roles, rather than to the roles themselves. The a priori argument then delivers the conclusion that conscious properties are identical to strictly physical properties. This is in fact how the argument is run by David Lewis.

a priori associations between concepts of conscious states and specifications of causal roles.

On this conceptual issue, I am thus in agreement with a number of recent writers who have argued that the a priori style of argument for materialism doesn't work (cf. Levine 1983, Chalmers 1996). They object to the initial a priori claim about concepts of conscious states. Our concepts of conscious states are not a priori related to any specifications of causal roles, they protest. So there is no conceptual route, they conclude, from the fact that any causal roles must be filled by physical states to the conclusion that conscious states are material.

I accept this criticism of the a priori style of argument for materialism. To repeat, I agree that our concepts of conscious states are not associated a priori with causal roles. But this isn't as bad for materialism as Levine and Chalmers suggest. If the a priori argument were the *only* argument for materialism about consciousness, then materialism would indeed be in trouble. However, it is not the only argument. There is also the original causal argument as I have presented it, which does not depend on any particular assumption about our concepts of conscious states.

1.10 *What is 'Physics'?*

Let me now address a terminological issue flagged earlier, an issue that may have been worrying readers for some time. How exactly is 'physics' to be understood in this context of the causal argument? An awkward dilemma may seem to face anyone trying to defend the crucial second premiss, the completeness of physics. If we take 'physics' to mean the subject-matter currently studied in departments of physics, discussed in physics journals, and so on, then it seems pretty obvious that physics is not complete. The track record of past attempts to list *all* the fundamental forces and particles responsible for physical effects is not good, and it seems highly likely that future physics will identify new categories of physical cause. On the other hand, if we mean by 'physics' the subject-matter of such future scientific theories, then we seem to be in no position to assess its completeness, since we don't yet know what it is.

This difficulty is more apparent than real. If you want to use the causal argument, it isn't crucial that you know exactly what a complete physics would include. Much more important is to know what it won't include. (Cf. Papineau and Spurrett 1999.)

Suppose, to illustrate the point, that we have a well-defined notion of the *mental* realm, identified via some distinctive way of picking out properties as mental. (Thus we might identify this realm as involving intentionality, say, or intelligence, or indeed as involving consciousness—the precise characterization won't matter for the point I am about to make.) Then one way of understanding 'physical' would simply be as 'non-mentally identifiable'—that is, as standing for properties which can be identified independently of this specifically mental conceptual apparatus. And then, provided we can be confident that the 'physical' in this sense is complete—that is, that every non-mentally identifiable effect is fully determined by *non-mentally identifiable* antecedents—then we can conclude that all mental states must be identical with (or realized by) something non-mentally identifiable (otherwise mental states couldn't have non-mentally identifiable effects).

This understanding of 'physical' as 'non-mentally identifiable' is of course a lot weaker than any normal pre-theoretical understanding, but note that it still generates a conclusion of great philosophical interest: namely, that all mental states, and in particular all conscious states, must be identical with non-mentally identifiable states. We may not know enough about physics to know exactly what a complete 'physics' might include. But as long as we are confident that, whatever it includes, it will have no ineliminable need for any distinctively mental categorizations, we can be confident that mental properties must be identical with (or realized by) certain non-mentally identifiable properties.

In fact, I shall understand 'physical' in a somewhat tighter sense in what follows, as 'identifiable non-mentally-*and*-non-biologically', or 'inanimate' for short, rather than simply as 'non-mentally identifiable'. This is because it is this realm, the 'inanimate', that is most naturally argued to be complete. When I examine the detailed scientific reasons for believing in the completeness of physics, in the Appendix, it will turn out that the realm which science has in fact shown to be causally sufficient unto itself is the inanimate. What science has actually shown is that any inanimate effect (that is, any effect specifiable in terms of mass, or charge, or chemical structure, or . . . in any non-biological and non-mental way) will have an inanimate cause. So it is this thesis that I propose to plug into the causal argument. Conscious causes have inanimate effects. Inanimate effects always have full inanimate causes. So conscious properties must be identical with (or realized by) inanimate properties.[12]

It might not be immediately obvious why I am being so careful here. Why not simply read 'physical' as non-mentally identifiable, as I suggested initially? If the Appendix succeeds is showing that the inanimate is complete, then won't it *a fortiori* show that the non-mentally identifiable is complete? After all, if something is inanimate, then it is certainly non-mentally identifiable. So, if the inanimate is complete, and there are inanimate causes for all inanimate effects, then those causes will be non-mentally identifiable too. And this would thus seem to ensure the completeness of the non-mentally identifiable.

No. This is too quick. To see why, take an *effect* which is *not* inanimate yet *is* non-mentally identifiable. An *arm* moving would be a good example. I take it that the notion of an arm movement is not a *mental* notion. But the notion of *an arm* is certainly a biological notion. So arm movements are not inanimate, even though they are non-mentally identifiable.

Now, the completeness of the inanimate tells us that all inanimate

[12] Note how 'animate' and 'physical $=_{df}$ inanimate' are not being used exclusively here. A property is 'animate' if it is identifiable pre-theoretically in mental or biological terms, as involving, say, intelligence, or consciousness, or respiration, or digestion. A property is 'inanimate' if it can be identified in some other way: paradigm 'inanimate' properties would thus be size, shape, mass, charge, and combinations thereof. This allows that certain properties can be both 'animate' and 'physical $=_{df}$ inanimate'. Such properties will be those that can be identified both ways: first, in 'animate' terms, and second, as (complexes of) 'inanimate' features which happen to be instantiated in minded or biological systems. This is of course the right way to set things up, given that we don't want definitions to rule out the materialist possibility that 'animate' properties are identical to 'inanimate' properties.

effects have inanimate causes. But, since *arm* movements aren't inanimate, it doesn't follow that they have inanimate causes, nor, therefore, that they must have non-mentally identifiable causes. Maybe, for all the completeness of the inanimate guarantees, arm movements are always caused by mental states alone, like desires or intentions, without any assistance from further causes at the inanimate level. This thus shows that the completeness of the inanimate doesn't guarantee the completeness of the non-mentally identifiable.[13]

This last point suggests a possible way of resisting the causal argument. Anti-materialists could allow that there is a familiar everyday sense in which conscious states have 'physical' effects, and another good sense in which the 'physical' realm is complete, and yet object that the causal argument fails to go through because the two senses are distinct. (Cf. Sturgeon 1998.[14]) Thus they could allow that conscious causes always have non-mentally identifiable bodily effects like arms moving, and also allow that the seriously inanimate realm of mass and motion is complete, but urge that since arm movements aren't themselves inanimate, it doesn't follow that they must have inanimate causes, nor therefore that their conscious antecedents must be identical with anything inanimately identifiable.

This is a serious enough issue, but it is scarcely conclusive against the materialist side. Materialists need only make sure that their senses of 'physical' line up properly. The version of completeness I

[13] The point is that switching from inanimate to non-mentally identifiable not only gives us fewer potential causes to falsify completeness, but also more effects, since the relevant completeness thesis now covers extra animate yet non-mentally identifiable effects, like raisings of arms. So we cannot infer from the fact that there are no *sui generis* mental causes for any inanimate effects that there are no *sui generis* mental causes for effects like the raising of arms. I would like to thank Finn Spicer for helping me to see this clearly.

[14] In fact, Sturgeon's charge is not that conscious causes fail to have inanimate effects, but rather that they fail to have quantum-mechanical effects. This is in line with his assumption that the materialists' causal argument will appeal to the causal completeness of quantum mechanics, rather than the causal completeness of the inanimate. However, as the Appendix will make clear, this isn't my completeness thesis. I don't think of quantum mechanics *per se* as asserting completeness, since the basic assumptions of quantum mechanics leave it open what forces (Hamiltonians) there are. Rather, my crucial completeness claim is that all inanimate accelerations are due to inanimate forces.

take to be defensible, as I said, is the completeness of the inanimate. So all I need to make the causal argument go through is a version of premiss 1 which will ensure that conscious causes do have inanimate effects, in addition to their effects on animate body parts.

One way of arguing for this premiss would be to start with the point that the anti-materialist concedes, namely, that conscious states cause animate effects, like arms moving, and then argue that arm movements should themselves be identitified with inanimate occurrences, thus giving the conscious causes inanimate effects, as desired. An obvious strategy here would be to note that arm movements themselves have inanimate effects (such as stones flying through the air, say), and then apply the causal argument once more, to conclude that these arm movements must be identified with the inanimate causes of those inanimate effects, if we are to avoid overdetermination.

But this is a somewhat long way round. If the animate bodily effects of conscious causes have inanimate effects, then we can infer directly that the conscious causes must themselves have those inanimate effects, by transitivity, whether or not the animate bodily movements are identified with inanimate occurrences. (Cf. Witmer 2000.) Thus, if some conscious desire causes my arm to move, and this movement in turn has such inanimate effects as a stone flying through the air, a window shattering, and so forth, then my conscious desire itself will cause these inanimate effects. I take it to be uncontroversial that conscious states standardly[15] have such inanimate effects, and will assume this henceforth.

1.11 *The Completeness of Physics*

Let me conclude this chapter with a few remarks about the causal argument's second premiss, the completeness of physics. It is one

[15] Mightn't *some* conscious occurrences, like wishful thinking, lack any inanimate effects? Well, I am happy to agree that the causal argument does not engage directly with such inefficacious states, and that this thus creates space for the possible view that specifically those conscious states are immaterial, while those with physical effects are not. But this does not seem a serious position to me. If anybody really wants to pursue it, they can send me an e-mail, and I'll think of some arguments.

thing to fix a sense of 'physics' which renders this a substantial claim which might be true or false. It is another to show that it is in fact true.

Some readers might feel that this is not a problematic issue. Once we have fixed a definite meaning for 'physical', as equivalent to 'inanimate', say, then is it not just a matter of common sense that all physical effects will have physical causes? In particular, if we take the physical effects in this sense that we normally attribute to conscious causes, then is it not obvious that these effects can always in principle be fully accounted for in terms of uncontroversially physical histories, involving the movement of matter (in arms), molecular processes (in muscles), the action of neurotransmitters (in brains) . . . and so on?

This is certainly how I thought of the issue when I first started working on the causal argument. I realized that this argument involved a number of disputable moves, and was therefore ready for it to be queried on various different grounds. But the one assumption that I did expect to be uncontroversial was the completeness of physics. To my surprise, I discovered that a number of my philosophical colleagues didn't agree. They didn't see why some physical occurrences, in our brains perhaps, shouldn't have irreducibly conscious causes.

My first reaction to this suggestion was that it betrayed an insufficient understanding of modern physics. Surely, I felt, the completeness premiss is simply part of standard physical theory. However, when my objectors pressed me, not unreasonably, to show them where the completeness of physics is written down in the physics textbooks, I found myself in some embarrassment. Once I was forced to defend it, I realized that the completeness of physics is by no means self-evident. Indeed, further research has led me to realize that, far from being self-evident, it is an issue on which the post-Galilean scientific tradition has changed its mind several times. The completeness of physics may seem the merest part of common sense to many of us today, but as recently as 150 years ago most people, including most orthodox scientists, would have thought the idea absurd, taking it to be obvious that there must be some *sui generis* conscious states in the causal history of human behaviour.

So the completeness of physics is a doctrine with a history, and a very interesting history at that. In the Appendix I detail this history. My main purpose in doing this is to show that there is good empirical evidence for the completeness of physics. But the historical story also shows that this evidence is relatively recent, and that prior to the twentieth century the empirical case for the completeness of physics was by no means persuasive.

At the beginning of this chapter I raised the question of why philosophical materialism has become popular only in the last fifty years or so. As I pointed out, this historical circumstance lends weight to the suggestion that contemporary materialism is a creature of fashion rather than serious philosophical argument. I take the story I tell in the Appendix to rebut this suggestion. There is indeed a good case for materialism. But it has not always been available to philosophers. This is because its crucial premiss, the completeness of physics, rests on empirical evidence which has emerged only relatively recently.

Chapter 2
CONCEPTUAL DUALISM

2.1 *Introduction*

The last chapter offered an argument for a materialist view of consciousness, where materialism is to be understood as a matter of property identity. Conscious properties are identical to material properties—that is, they are identical either to strictly physical properties, or to physically realized higher properties.

Still, while I am a materialist about conscious *properties*, I am a sort of dualist about the *concepts* we use to refer to these properties.[1] I think that we have two quite different ways of thinking about conscious properties. Moreover, I think that it is crucially important for materialists to realize that conscious properties can be referred to in these two different ways. Materialists who do not acknowledge this—and there are some—will find themselves unable to answer some standard anti-materialist challenges.

I shall call these two kinds of concepts 'phenomenal' concepts and 'material' concepts. I shall have plenty to say about both kinds of concepts in what follows. But it will be helpful to start with a rough initial characterization.

[1] In *Remnants of Meaning* Stephen Schiffer (1987) combines a 'sentential dualism' with an ontological physicalism about propositional attitudes. There are affinities between this and my 'conceptual dualism' about conscious experiences, though also many specific differences.

Material concepts are those which pick out conscious properties *as* items in the third-personal, causal world. Most commonly, these will be role concepts, by which I mean concepts which refer by describing some causal or other role, such as pain's role in mediating between bodily damage and avoidance behaviour.[2] But I want also to include under this heading directly physical concepts which identify their referents in terms of some intrinsic physical constitution—for example, in terms of shape, mass, charge, and so on.[3]

The category of phenomenal concepts is less familiar. The general idea is that when we use phenomenal concepts, we think of mental properties, not as items in the material world, but in terms of *what they are like*. Consider what happens when the dentist's drill slips and hits the nerve in your tooth. You can think of this materially, in terms of nerve messages, brain activity, bodily flinching, facial grimaces, and so on. Or you can think of it in terms of what it would be *like*, of how it would *feel* if it happened to you.[4]

Now, as a materialist, I hold that even phenomenal concepts refer to material *properties*. In distinguishing phenomenal concepts from material concepts, I do not wish to suggest that they refer to different entities. The argument of the last chapter gave us every reason to take the two kinds of concepts to make common reference to material properties. The idea, then, is that we have two quite different ways of thinking about *pain*, say, or *tasting chocolate*, or *seeing an elephant*,

[2] As we saw earlier (Ch. 1 n. 11), role concepts of properties can be of two types. They can name whichever property realizes the role, or they can name the higher property which constitutes the role. This distinction won't matter in the present chapter. (It is perhaps worth noting that it is not a priori, even if it is true, that material role *concepts* of either kind should name material *properties* as defined in the last chapter; for it is not a priori, even if true, that the relevant roles will be physically realized.)

[3] I could perhaps also have included some *perceptual* concepts under the heading of material concepts of conscious states, such as visual concepts of certain brain states. But since such perceptual concepts of brain states play no prominent role in my arguments until Chapter 6, it will simplify things to leave them out.

[4] My distinction between 'phenomenal' and 'material' concepts is similar to David Chalmers's distinction between 'phenomenal' and 'psychological' concepts (Chalmers 1996). But his 'psychological' concepts are specifically role concepts, and for present purposes it is more convenient to work with my more general category of non-phenomenal 'material' concepts. In Chapter 4, however, I shall have occasion to make use of Chalmers's 'psychological' category.

both of which refer to the same material properties in reality. By way of an obvious analogy, consider the case where we have two terms, 'water' and 'H_2O', say, both of which refer to the same liquid.

We might say that the difference between phenomenal and material concepts is a difference at the level of sense, not reference. As in standard cases of co-reference, we have two terms which refer to the same entity, but in different ways—that is, in virtue of different senses. There will be many questions to answer about these distinct modes of reference, and in particular about the mode in which phenomenal concepts refer. But the underlying assumption will remain, that these different modes both point to the same objective material property.[5]

If phenomenal and material concepts are quite distinct at the level of sense, there will be no a priori route to the identification of their referents. Examinations of the concepts themselves will not tell us that they refer to the same properties. Such knowledge can only be arrived at a posteriori, on the basis of empirical evidence about their actual referents. Still, this will not worry materialists who defend materialism in the way outlined in the last chapter. For nothing in that line of argument depended on any a priori analysis of concepts.

Ned Block (forthcoming) has recently coined some useful terminology. He uses the term 'inflationists' for philosophers who recognize an extra range of phenomenal concepts. Not all materialists are inflationists. As we shall see, a number of leading materialist philosophers, including David Lewis and Daniel Dennett, deny phenomenal concepts, and hold that all references to conscious states are made using material concepts alone. Since these philosophers do not recognize any distinctive conceptual

[5] To forestall one possible confusion, let me make clear that I do not take my conceptual 'dualism' itself to demand any special non-material ontology. In my view, the deployment of phenomenal concepts depends on material processes in thinkers' brains, just as much as the deployment of any other concepts. Indeed, I shall shortly say something more about the brain processes that might underly the deployment of phenomenal concepts. It is true that the ontology of concepts themselves is a somewhat obscure matter. Some philosophers would argue that they are a species of abstract entity, akin to numbers. I have my doubts about this, and would hope to parse away any such references to concepts as abstract objects. But, in any case, questions to do with abstract objects are quite independent of any of the issues addressed in this book.

apparatus for referring to conscious states, Block calls them *'deflationists'*.

2.2 *Jackson's Knowledge Argument*

The best way to demonstrate the existence of phenomenal concepts is via Frank Jackson's 'knowledge argument' (1982, 1986). Jackson himself originally proposed this argument as a way of demonstrating the existence of distinctive phenomenal *properties*—that is, conscious properties which cannot be identified with any material properties, and which therefore refute materialism. I think that his story does not establish this anti-materialist conclusion, and will shortly argue as much. But at the same time it does provide an excellent way of establishing the existence of distinctive phenomenal *concepts*.

Jackson's argument is made graphic by his well-known 'Mary' thought-experiment. Mary is some future cognitive scientist. She is an absolute authority on human vision, and in particular on colour perception. She has complete material knowledge about what goes on in humans when they see colours. She knows all about light waves, and reflectance profiles, and rods and cones, and about the many areas concerned with vision in the occipital lobe, and what they each do, and about the kinds of circumstances that produce different colour experiences, and the kinds of illumination that produce colour illusions, and so on.

However, apart from this, Mary has had a somewhat unusual upbringing. She has never seen any colours herself. She has lived all her life inside a house painted black and white and shades of grey. All her knowledge of colour vision is book learnin', and none of her books contains any colour illustrations. She has a TV, but it is an old black-and-white set.

Then one day Mary walks out of her front door, and sees a red rose. At this point, Jackson observes, Mary learns something new, something she didn't know before. She learns what it is like to see something red.

Jackson takes this to show that Mary becomes acquainted with some new *property* of red experiences, the 'conscious feel' of red experience.[6] After all, before she came out of the house, she already knew about every material property of red experiences. If she learns about something new, argues Jackson, this must involve her now knowing about some *further* feature of red experiences, the conscious feature, which cannot therefore be identical with anything material.

However, materialists who recognize phenomenal concepts needn't accept this argument. They can respond that, while there is indeed a genuine before–after difference in Mary, this is just a matter of her coming to think in new ways, and in particular of her acquiring a new *concept* of seeing something red. There are no new experiential properties in the offing. The property she refers to with this concept is still a perfectly good material property, that material property, whatever it is, that is present in just those people who are seeing something red, and which she could think *about* perfectly well, albeit only using material concepts, even before she saw the rose.

2.3 *Denying Any Difference*

Let me go a little more slowly. Not all materialist philosophers respond to Jackson's argument in this way—that is, by arguing that Mary is changed at the level of concepts, even if not by any acquaintance with new phenomenal properties. I shall consider two alternative materialist responses which deny that she acquires any new concepts. These are 'deflationist' responses to Jackson's argument, in that they see no reason to credit Mary with anything but material concepts, even after she leaves her house. Exposing the deficiencies in these deflationist strategies will help to make it clear why materialists need to recognize distinctively phenomenal concepts.

[6] In calling experiences 'red', I do not of course mean that they have the same quality as ripe tomatoes or pillar-boxes. A more careful phrase would be 'an experience as of seeing something red'. But I shall use the less careful 'red experience' when it is expositorily smoother to do so, and in no danger of causing confusion. I shall also sometimes ease the exposition by using 'seeing something red' for the longer 'an experience as of seeing something red'; in general this should not be taken to imply that the thing seen is actually red, or even that something actual is seen.

The first deflationist strategy, which is most prominently defended by Daniel Dennett (1991), aims to stop the Mary argument before it starts, by denying that Mary displays any significant before-after difference in the first place. This strategy will be addressed in this section, and will lead, in the following two sections, to an initial explanation of why Jackson's argument fails to establish ontological dualism. The second deflationist strategy, widely known as the 'ability hypothesis', allows that Mary is significantly changed when her new experience shows her 'what seeing something red is like', but insists that this change involves her acquiring only new abilities, not new concepts. This strategy will be explained in section 2.6 below.

As I said, the Dennettian strategy denies that Mary undergoes any substantial change in the first place. Of course, there is one trivial before-after difference, which can be agreed on all sides. This is that Mary has a new experience after she comes out of the house, an experience of a kind she has never had before. This is not at issue, for there is nothing in this to provide any argument against materialism. Materialists are just as well placed as anybody else to explain this difference. Materialists think that conscious experiences are identical with certain material occurrences in the brain. So materialists can simply say that *this* before-after difference in Mary, that she has now had an experience which she hadn't had before, is simply that certain material states—namely, those which constitute red experiences—have now occurred in her, when before they hadn't.

The more important question is whether there are any *further* before-after differences in Mary, consequent on her having had this experience. Jackson wants to say that, in addition to having had the experience, she now also *knows* something she didn't know before: namely, what the experience is like. This knowledge isn't just a matter of once having had the experience itself. It is something that remains with Mary after the experience is over. With luck, she'll now retain her knowledge of what seeing something red is like throughout her life. It is this further change that Jackson takes to present a problem for materialists. Since Mary knew about all the material properties of red experiences before she came out of the house, argues Jackson, her *extra piece of knowledge* means that red experiences must have some non-material property.

Dennett (1991) seeks to block this argument by denying that Mary will undergo any change of the kind Jackson supposes. According to Dennett, Mary won't in any sense learn anything new when she comes out of the house. Whatever we understand by 'knowing what it is like', argues Dennett, Mary already knows what it is like to see red.

Dennett allows that ordinary people learn something new from new experiences. But Mary is no ordinary person. She is supposed already to know *everything* material about colour experience. Dennett argues that this removes her so far from the familiar that we should not trust our intuitions about her, and in particular our intuition that she will learn something new from her experience. Ordinary people may learn from experience. But Mary already has absolutely complete information, and so has nothing left to learn.

Or at least so Dennett argues. However, this line of argument seems quite implausible. Dennett is looking in the wrong place for the relevant before–after differences. The important changes occasioned in Mary, and signalled by the phrase 'she now knows what it is like', should not be thought of as her somehow expanding her stock of ordinary material knowledge. Indeed, she can't do this, by hypothesis. But there remains the possibility, to which Dennett seems blind, that Mary will instead acquire some quite new powers of thought, of a kind she simply didn't have before.

The important point, which I think even materialists should concede to Jackson, is that Mary's new experience will enable her henceforth to *re-create* this experience in *imagination*, and in addition to *classify* new experiences *introspectively* as of the same kind. This is the most natural way of reading the expression 'coming to know what something is like'. Mary is changed, not through getting more knowledge of the material kind she previously had, but through aquiring these two new powers of imagination and introspection.

Thus, someone who undergoes a new kind of experience will later be able to imagine what the experience is like, in a way they couldn't before. They will have a grasp of the *redness* of red experience, so to speak. In addition, someone who undergoes a new kind of experience will thenceforth be able introspectively to categorize further experiences as feeling like *that*. They will be able directly to pick out an aspect of current experiences as manisfesting that characteristic redness.

Now, the analysis of these imaginative and introspective powers will occupy much of the rest of this book. But at this stage, even before we go into the details, we can see what is unconvincing about Dennett's line that the housebound Mary will already 'know what seeing something red is like'. If 'knowing what it is like' is read along the lines suggested above, Dennett would seem to be committed to the view that the pre-experiential Mary, in virtue of her encyclopaedic knowledge, can already imagine what it is like to see something red, and is already poised to classify further experiences directly and introspectively as of that type.

This seems wrong. In the next two sections I shall offer a natural explanation of *why* Mary can't do these things prior to her own red experience. But, even prior to this explanation, it seems clear *that* she won't be able to do these things before she emerges from her house. No amount of book learnin' will tell her how to create the experience of red in imagination, or how introspectively to classify further experiences as of that type. To suppose otherwise is to suppose that such non-experiential learning on its own will somehow enable you to enact a red experience imaginatively, and show you how to judge introspectively whether or not some further experience involves that feeling. This would surely be very weird.

For any readers who may remain unconvinced, I need not press the point at this stage. The rest of this chapter will make it amply clear why Dennett's line is both unnatural and unnecessary for a materialist. The reason why Dennett himself takes this line, I suspect, is that he is strongly committed to some kind of 'deflationist' analysis of concepts of mental states. He assumes that there is no other respectable way of thinking about mental states apart from thinking of them in terms of roles—that is, *as* states with certain canonical links to behaviour and perhaps other similarly identified mental states. And of course, if you do take this to be the only respectable way to think about mental states, then you must conclude that Mary's new experience couldn't possibly lead to any new information, since she already had all the information that could possibly be framed using such material role concepts of mental states.

Still, it seems desperate to end up denying, as does Dennett, that there is no real before–after difference in Mary. Surely even materialists should admit that Mary is changed in some lasting way.

The question which then faces materialists is whether this change amounts to Mary acquiring a new concept, as conceptual inflationists like myself want to assert, or whether the change can still be understood in a conceptually non-inflationary way. In section 2.6 I shall consider a version of materialism which goes beyond Dennett in allowing a real before–after difference in Mary, yet aims to stop short of the inflationist view that this is a matter of her gaining some new concept. This is the version of materialism known as the 'ability hypothesis'.

However, before we consider this 'ability hypothesis', we need to look more carefully at the before–after differences in Mary. Now that we are agreeing, contra Dennett, that there are substantial before–after differences, we had better make sure that they do not imply some new acquaintance with phenomenal *properties*, in the way Jackson supposes.

I said that the crucial differences lie in Mary's new powers of *imaginative re-creation* and *introspective classification*. In the next two sections I shall accordingly consider these in turn, showing in each case that there is no legitimate argument from before–after difference to distinct phenomenal properties. The Mary argument does not establish any dualism of properties.

Then I shall turn to the 'ability hypothesis'. This upholds the basic materialist line that Jackson's argument does not imply any dualism of properties. But it does allow, contra Dennett, that Mary acquires new powers of imaginative re-creation and introspective classification. However, it also seeks to distance itself from any dualism of concepts—that is, from the inflationist claim that Mary acquires a new phenomenal concept. I shall show that even this more sophisticated form of deflationism fails to deal adequately with Jackson's argument, and that the story of Mary leaves us with no alternative but to admit distinctive phenomenal concepts.

2.4 *Imaginative Re-creation*

The first before–after change to be considered concerns Mary's new powers of *imaginative re-creation*. Once she has seen red, Mary can re-create the experience of seeing something red, whereas before she couldn't. Mary could of course always imagine, in the third person, so to speak, that somebody else was seeing something red, in the sense that she could entertain the possibility of such-and-such material occurrences in another person. But now she has a new ability. She is able to imagine *having* the experience itself, from the inside, as it were. She can now *relive* the experience, as opposed to just thinking about it.

Anti-materialists like Jackson (1982, 1986) want to account for this change in terms of Mary's new acquaintance with some non-material property. The anti-materialist story would go something like this. When Mary experiences red, she becomes acquainted with the characteristic phenomenal feature of red experiences. And henceforth this acquaintance enables her to imagine the experience in question, since she will now be able to call this property to mind, and thereby re-create in her mind the characteristic phenomenal feel of red experiences.

However, there is an obvious alternative materialist story to be told. This accounts equally well for the fact that you can't imagine an experience prior to having it, and does so without invoking any special phenomenal properties.

Here is the obvious materialist explanation. Suppose that imaginative re-creation depends on the ability to reactivate some of the same parts of the brain as are activated by the original experience itself. Then it would scarcely be surprising that we can only do this with respect to types of experience we have had previously. We can't form replicas, so to speak, if external stimulation hasn't fixed a mould in our brains. Less metaphorically, we can only reactivate the parts of the brain required for the imaginative re-creation of some type of experience, if some actual experience of that type has previously activated those parts.[7]

[7] Note in particular how this materialist story avoids any dubious idea of direct aquaintance with such phenomenal properties. I find the anti-materialist story

There is now plenty of evidence to support this hypothesis about imaginative re-creation. Data from brain scans and similar techniques show directly that imagination activates some of the same parts of our brains as are activated by actual experiences of the relevant type. Moreover, studies of patients with brain lesions shows that damage to the relevant areas can also destroy imaginative abilities. People with damage to certain parts of the visual cortex will lose the ability not only to see, but also to visually imagine. Both these lines of evidence strongly suggest that imaginative re-creation is a matter of 'turning on' some characteristic pattern of brain activity that was first created by an original experience.

These remarks merely gesture at a complex body of empirical data. But they suffice to indicate how materialists might explain Mary's new power of imaginative re-creation, yet deny that it demands any new non-material property. Mary's new power does not depend on any acquaintance with such a phenomenal property. Rather, her brain is lastingly altered in certain ways, and this now allows her imaginatively to re-create an experience that she could previously only think about materially. Seen in this way, it is clear that there is nothing in the idea of imaginative re-creation to worry materialists.

2.5 *Introspective Classification*

The other change in Mary was to do with introspective classification. Once she has seen red, she can introspectively classify further experiences as of that type. We can think of Mary as acquiring a new classificatory category, for which she might have no word, but which she can apply to particular new experiences.

Anti-materialists will again maintain that this new power of introspective classification testifies to Mary's direct acquaintance

especially puzzling at this point. In particular, how is the before–after change in Mary supposed to be sustained after she stops having her new experience? Can she now reproduce the phenomenal property in her mind at will, so as to *reacquaint* herself with it? Or can her memory reach back through time to keep *acquainting* her with the earlier instance? Both these ideas seem odd, but something along these lines seems to be needed to explain why Mary is *permanently* changed by her experience.

with some distinct phenomenal property. Before she experienced red, she had never been in contact with this phenomenal property. But now that she is acquainted with it, she can classify new experiences according to whether they display it or not.[8]

Once more, however, there is an obvious materialist story to be set against this, which accounts equally well for the fact that people can introspectively classify only into experiential kinds that they have themselves previously instantiated, yet does so without invoking any special phenomenal properties.

Suppose that introspective classification depends on the existence of some kind of brain 'template', to use David Lewis's phrase (1983). We don't classify new experiences by seeing whether they have some phenomenal property with which we have previously been acquainted. Instead, we simply compare them with the 'template' to see whether they correspond. This hypothesis too yields an obvious materialist explanation of why you should only be able to introspectively classify experiences of a kind that you have previously had. Again, the brain needs an original to form the mould. In order to fix a neural pattern as a template against which to compare new inputs, we need some original experience to create the pattern.

To make this template hypothesis more concrete, we might suppose that, whenever the relevant classificatory question arises, the template 'sends down' neural signals to lower levels of perceptual processing. A positive classification would then be triggered by a 'match' between these backwards signals and current sensory input. This match could then boost the activation of the template, and this boosted activation could itself serve as the relevant classification of current experience.

Now, the precise accuracy of this picture is clearly hostage to empirical research. Still, if anything even roughly along these lines is right, it will yield a natural materialist explanation of why Mary's

[8] Again, there are puzzles about the role of direct acquaintance in the non-materialist story. How does it help in classifying new experiences that you were *previously* acquainted with some phenomenal property? The questions pressed about imaginative re-creation in n. 7 above apply equally to introspective classification.

experience should enable her to think in ways she couldn't think before—moreover, an explanation which doesn't require any distinct phenomenal properties.

2.6 *The Ability Hypothesis*

My overall aim in this chapter is to draw an inflationist moral from the Mary thought-experiment. When Mary comes to 'know what seeing something red is like', she acquires a new kind of concept of seeing something red, a phenomenal concept, which is quite different from any material concepts she previously possessed. She mightn't be acquainted with any new phenomenal *property*—some of her old material concepts already referred to the property of seeing something red—but she has a new way of thinking about that property.

It might not yet be clear, however, what justifies counting the before–after changes in Mary as amounting to her acquisition of a new *concept*. We have seen how she will have new powers to imaginatively re-create and introspectively classify red experiences. But why view these changes as the acquisition of a concept?

However, note that Mary's new powers apparently enable her to think certain new kinds of thought. Now that she can imagine red experiences, Mary can think thoughts like 'People looking at ripe tomatoes experience *this*'. And her new introspective powers will also allow her to think thoughts like '*This* is what people experience when they look at ripe tomatoes'. That is, she will be able to deploy her imaginative and introspective powers in the construction of articulated thoughts, mental judgements that can be true or false.

When I speak of 'concepts', I mean components in just such truth-evaluable thoughts. Concepts are elements which make a systematic contribution to the truth conditions of the thoughts they enter into. By this criterion, the 'this's in the thoughts attributed to Mary above would seem to represent something conceptual. For the overall thoughts containing the 'this's certainly seem to be evaluable as true or false.

Some philosophers are happy to accept that Mary acquires new powers of imaginative re-creation and introspective classification, yet deny that it is appropriate to view this as a matter of her acquiring any new phenomenal *concepts*. These are the sophisticated deflationists of the 'ability hypothesis' mentioned earlier. They accept, contra Dennett, that some genuine and lasting before–after differences are occasioned in Mary by her new experience. In particular, they accept that she will now be able to re-create that kind of experience in imagination, and to classify new experiences introspectively as of that kind. Yet they deny that Mary will thereby acquire any new concepts. All she acquires are some new imaginative and introspective *abilities*. She can perform imaginative and introspective acts which she could not perform before (Lewis 1988, Nemirow 1990).

It may seem that, in conceding this substantial before–after difference in Mary, defenders of the ability hypothesis have no option but to concede that she acquires a new concept. If she comes to *know* something she didn't know before—she comes to 'know what seeing something red is like'—then doesn't it immediately follow that she must have some new thoughts, at least in the sense that her thoughts involve new concepts, even if they refer to items she could always refer to?[9]

Not necessarily, according to the ability hypothesis. For knowledge can include knowledge *how*, as well as knowledge *that*. When I find out how to ride a bicycle, I come to know something I didn't know before. I now know how to ride a bicycle. But this needn't involve me having any new thoughts—it needn't involve me knowing *that* anything I didn't know before. After all, I may already have known everything of a propositional kind about riding bicycles, even before

[9] Given that the issue at hand hinges on whether Mary is capable of 'new' thoughts, we need to take care about typing thoughts. After all, in one sense even inflationists like myself want to deny that Mary can think new thoughts: she doesn't think *about* any new entities, she only refers to the same old entities using new concepts. So, if we type thoughts at the level of reference, then even inflationists deny that Mary has any new thoughts. Still, if we type at the level of sense, then inflationists hold that Mary *is* thinking new thoughts, in that her thoughts deploy new concepts. The sophisticated deflationist defenders of the ability hypothesis want to deny even this much: they claim that even at the level of sense or concepts, Mary doesn't have any new thoughts.

I learned how to ride one. I could have been an absolute expert on the physics, physiology, economics, and history of bicycle riding, and just not have acquired the knack myself. If so, what I would have lacked was not any kind of *thoughts* about bicycle riding, however typed, but simply the ability to ride a bicycle myself.

So it is with Mary, according to the ability hypothesis. In a sense, she did not 'know what seeing something red is like' before she came out of her house. But she wasn't in any way incapable of thinking thoughts about red experience. All she lacked was the ability to re-create that experience in imagination and the ability to classify it by introspection.

As a version of deflationism, the ability hypothesis is clearly preferable to Dennett's outright denial that Mary in any sense comes to 'know something new'. It allows that Mary acquires new powers of imagination and introspection. Even so, the ability hypothesis does not really do justice to the change in Mary. If we look more closely at Mary's new abilities, we will see that they are inseparable from her power to think certain new kinds of thoughts.

Go back to the examples at the beginning of this section. Mary imagines a red experience and thinks 'People looking at ripe tomatoes experience *this*'. Or she introspects her current experience, and thinks '*This* is what people experience when they look at ripe tomatoes'. The ability hypothesis needs to argue that this mode of expression is misleading. Since Mary has no new concepts, her thoughts can involve only her old material concept of red experience. Rather, the novelty in her thinking, such as it is, lies in her having new *routes* to these thoughts, not in their containing new concepts.

To take introspective classification first, the idea would be that Mary has a new *introspective route* to beliefs involving her old material concept of red experience. She can now arrive at beliefs with this content directly and introspectively, whereas before she could ascribe red experiences to people only on different grounds, by knowing about their behaviour, say, or by knowing about physical goings-on inside them. But there isn't any new concept here. She is just using the old material concept of red experience she had before she left the house. She's just acquired a new technique for applying the concept.

And a similar line might be taken with the thoughts involved in Mary's new imaginative abilities. Mary re-creates the experience of seeing something red in her imagination, and simultaneously thinks something like 'Ripe tomatoes cause this'. But perhaps the concept expressed here by 'this' is simply identical to the old material concept of red experience that Mary always possessed. Her thought involving this concept is now accompanied by an act of imaginative re-creation, an act that Mary couldn't have performed before she had her first red experience. But the thought itself involves only concepts she always had available.[10]

Once spelt out, the problem with this line is obvious. Why suppose that the concepts involved in Mary's introspective and imaginative thoughts can be equated with the old material concepts she always possessed? When I imagine seeing something red, and think '*This* is caused by ripe tomatoes', I certainly don't seem to be deploying any particular material concept. Nor do I seem to be doing so when I introspectively think '*this* is caused by ripe tomatoes'.

To drive the point home, note that Mary may not yet know *which* of her old material concepts applies to her new experience. Imagine that she is shown, not a rose, but a coloured sheet of paper, so she has no way of knowing, in her old material terms, which *colour* experience this is. She might be able to figure out that it is a colour experience, but there is nothing to tell her whether she is seeing something red or green or blue. This shows that Mary cannot be thinking just using her old material concepts.

Suppose that Mary, after being shown the piece of red paper, uses her new imaginative powers to hazard 'I'll have *this* experience again before the day is out'. This is clearly a thought in full working order—after all, it will either be true or false—but equally clearly it is not equivalent to any thought that Mary can form using her old material concepts, since she has no idea which of those picks out *this*

[10] It is not entirely clear that Lewis (1988) is committed to this deflationist denial that Mary acquires a new concept. Lewis does allow that Mary may acquire a new 'word' in her 'language of thought'. But he also compares this to adding Russian words to English ones—which needn't of course involve acquiring new concepts. Perhaps Lewis's caginess about inflationism stems from the fact that his argument for physicalism, elaborated in Lewis 1966 and elsewhere, requires that our concepts of mental states are non-phenomenal role concepts.

experience. Again, suppose that Mary later confirms her guess with the introspective classification 'Aha—*this* is the experience I first had this morning'. As before, this claim could be true or false, yet it can't possibly be equivalent to any claim made using one of Mary's old material concepts. since again she will not know which of these picks out *this* experience.[11]

There is a sense in which Mary's new powers of imaginative re-creation and introspective classification are indeed new abilities—she can certainly do things she could not do before. But they are not *mere* abilities, if that is taken to rule out her possession of new phenomenal concepts. At the level of reference, Mary may still be thinking about the same properties she could always think about. But at the level of sense, her new imaginative and intropective powers generate a new way for her to think about those properties.

Let me conclude this section by drawing attention to a feature of phenomenal concepts that has so far been left implicit. This is the fact that phenomenal concepts can refer both to particular experiences and to types of experience. This ability, to refer to both particulars and types, is displayed by other kinds of general concepts. (Thus we can say both that 'The electron is attached to the oil drop' and that 'The electron has negative charge'; or again, 'The whale has escaped' or 'The whale is a mammal'.) Phenomenal concepts are similar in this respect. Thus Mary might use imaginative re-creation to think about a type, as in 'That experience was very exciting—I hope I have it again'—or, alternatively, to think about a particular experience, as in 'That experience must have been caused by what I ate last night'. And the same contrast will be present in thoughts grounded in introspective classification. Thus, 'I wouldn't mind having this experience more often' versus 'This experience can't last much longer'.

2.7 *Indexicality and Phenomenal Concepts*

Some of the phrases I have been using to express phenomenal concepts may have suggested that phenomenal concepts are a

[11] For similar arguments against the ability hypothesis and in favour of conceptual dualism, see Loar 1990.

species of indexical concept. I have typically alluded to both imaginative and introspective uses of such concepts with the construction '*this* experience'. Given this, perhaps we can explain the workings of phenomenal concepts in terms of indexical constructions.

In Chapter 4 I shall examine this idea at some length, and offer a fairly detailed account of how far phenomenal concepts do and do not resemble familiar indexical constructions. But in this section I want to make one preliminary point: any account of phenomenal concepts in terms of indexical constructions must respect the fact that phenomenal concepts make essential use of powers of imaginative re-creation and introspective classification.

In arguing for this point, I intend to rule out an alternative and rather more straightforward indexical account of phenomenal concepts. On this alternative account, the reason why Mary acquires a new way of referring to red experiences has nothing to do with her having any new imaginative or introspective powers. Rather, it is simply a matter of her now being able to ostend one of her past red experiences, when previously she couldn't, for lack of any such experiences. Her new concept is thus simply 'that experience', where the 'that' points indexically to some past red experience. Of course, Mary could always form concepts of roughly this kind by ostending *other* people's experiences, even before she came out of her house. But what is indeed new for Mary is her ability to do this by invoking one of her *own* past experiences. Whereas previously she could only refer indexically to experiences at second hand, now she can do it at first hand.

So, on this suggestion, Mary's acquisition of phenomenal concepts is simply a matter of her now being able to ostend some past experience of her own, and thereby form a term that refers to that experience. If this were right, imaginative re-creation and introspective classification would play no special role in forming phenomenal concepts. All you need is some past instance of the experience in question.

As a preliminary to showing why this alternative suggestion doesn't work, let me first sketch a crude model of indexical constructions in general. Let us take it that indexical terms always involve a demonstrative element ('this', 'that', or perhaps simply pointing) plus a descriptive element ('animal', 'shape', 'car'). The compound indexical term ('that animal') then refers to the unique entity, if there is one such, that both lies in the 'direction' indicated by the demonstrative element and satisfies the descriptive term. Some indexical phrases run together both demonstrative and descriptive components ('now' = 'this *time*', 'there' = 'that *place*'), but this terminological fact does not affect the underlying model.

We can think of the descriptive element in an indexical construction as fixing some range of possible referents, and the demonstrative element *plus* the contents of the indicated 'direction' as then narrowing down this range to some specific referent. Note that it is consistent with this model that indexical constructions—just like phenomenal concepts and other general concepts—can be used to refer to both types and particulars. The phrase 'that car' can be used to pick out either a model (the Rolls-Royce Corniche, say) or some specific car (Tom Jones's Roller). The disambiguation here can be done explicitly ('that make of car') or left to the conversational context.

Let me now return to phenomenal concepts themselves. The suggestion to be examined is that Mary's distinctive new powers of reference lie solely in the fact that she is now in a position to demonstrate past experiential items of her own, where previously she could not. On this suggestion, the formation of phenomenal concepts owes nothing to acts of imaginative re-creation or introspective classification, but simply to the availability of past instances of the experience in question.

The sharpest way of showing that this suggestion does not work is to consider the kind of case where Mary uses a phenomenal concept to think about red experiences (the *type*, let us suppose, for the sake of specificity) after her original red experience is over. As I have been representing this case, Mary re-enacts her original experience in imagination, and therewith thinks about that experience. According to the suggestion under examination, however, the imaginative re-creation plays no essential role. Mary is simply demonstrating the relevant experiential type in question by pointing back in time to one of her own experiences.

But suppose that Mary has forgotten when and where she had this earlier experience, and has no other identifying information about it, and so cannot demonstrate it using standard indexical constructions. This won't stop her being able to refer to that type of experience as 'this experience', accompanied by some act of imagination. She will still be able to think thoughts like 'I'd love to have *this* experience again, even though I can't for the life of me remember when or where I previously had it'.

Exactly how the 'this' here interacts with the act of imaginative re-creation will be explored further in Chapter 4. For the moment my concern is only to establish that the imaginative act plays an essential role. Mary isn't simply pointing back in time to the occasion where she first had that kind of experience; she is somehow using her new power of imagination to pick out that kind.

It may seem as if the model I am disputing will work better when we turn to cases where thinkers refer to experiences they are currently undergoing. As I have been representing this kind of case, the thinker performs an act of introspective classification, and therewith thinks about the experience in question. But why suppose the act of introspective classification plays any essential role here? When I look into myself, and refer to some aspect of my experience as '*this* feeling', am I not simply pointing internally to something occurring inside me?

But even here there are difficulties. It seems unlikely that the indexical construction '*this* feeling' is ever *well directed* enough to identify some specific aspect of an individual's overall conscious experience. At any time an individual's conscious experience will be multi-faceted and multi-modal. You can see many different features and objects at a given time, not to mention further awareness involving hearing, smelling, itching, and so on. Given this, you will need something more than a simple 'this feeling', pointed generally inside yourself, so to speak, to determine which *aspect* of your current experience is being referred to.

This is where I take introspective classification to play an essential role. It serves to 'highlight' some specific aspect of your current experience, and thereby to render your phenomenal concept 'this experience' determinate. Without any such introspective classification, there would be nothing to fix which of the many features of your overall experiential state is being referred to. Simply pointing inwards to your current manifold experience could not possibly constitute a referential act, without introspective classification to focus reference on some specific feature.[12]

Again, there is more to be said about the way in which introspective classification enters into phenomenal concepts, and I shall return to this in Chapter 4. For the present I want only to establish that introspective classification plays some such role.

2.8 *The Contingency of Learning from Experience*

In some ways the inflationist account of phenomenal concepts is reminiscent of the traditional empiricist account of ideas. Hume maintained that all ideas are copies of impressions. You can only think with concepts that you have derived from earlier experiences. Inflationists need not agree with Hume that *all* concepts have such an experiential source. Inflationism is a thesis specifically about phenomenal concepts of conscious properties, and carries no implication about *non*-phenomenal concepts of conscious properties

[12] Might not certain *material* concepts of modes of experience fit the bill instead? I am thinking here of general material concepts like *vision*, *hearing*, or *smell*, or, at a more fine-grained level, *colour vision*, *pitch perception*, and so on. Couldn't we use these, instead of introspective classification, to pick out specific aspects of our current overall experience, say as '*this* smell', or as '*this* colour experience'? But this suggested explication of introspective phenomenal thinking can be discredited by a further variant of the Mary thought-experiment. Consider a Mary who has never *heard* anything or *smelled* anything. Then she has a new experience. She hears middle C, say. However, we set things up so that she doesn't know, in her old material terms, whether this is a sound or a smell. So she won't know whether to pick out this experience, using her old material terms, as 'this sound' or as 'this smell'. Yet I take it that she will be able to refer in phenomenal terms to this or any further such experiences as '*this* experience'. This makes it clear that such phenomenal thinking uses Mary's newly acquired power of introspective classification to achieve referential determinacy, rather than any general material concepts that were always available to her.

or of anything else. But when it comes to these phenomenal concepts themselves, inflationists do maintain that you need an initial experience, as with Mary, to acquire a phenomenal concept.

Of course, some obvious qualifications are needed, analogous to those originally made by Hume about complex and intermediate experiences. It is possible to know what *complex* experiences are like—that is, to be able to re-create them imaginatively and to classify them introspectively—even if you have never had them before, provided you have had the *simple* experiences out of which these complex experiences are composed. For example, we are capable of imaginatively re-creating and introspectively classifying the complex experience of seeing a red circle, even if we've never seen one before, as long as we've previously experienced the elements of seeing a circle and seeing something red. In addition, it is arguable that we are sometimes capable of imaginatively creating or introspectively classifying *intermediate* experiences that we haven't had before, provided we have previously had relevantly related experiences. For example, we might sometimes be able to imagine or classify a colour experience which is spectrally intermediate between other colour experiences we have previously undergone.

It is not hard to think of materialist explanations for these qualifications to the Humean principle, and accordingly I shall take them as read from here on. Now, subject to these qualifications, it is worth asking *why* you can only 'know what an experience is like' once you have had it yourself. It is striking that materialism implies that this is a quite contingent matter, whereas the anti-materialist alternative suggests that it is necessary.

Thus Jackson's anti-materialist argument assumes that 'knowing what it is like' requires some kind of acquaintance with a non-material property. On this view it is therefore a matter of necessity that you must undergo an original experience before you can acquire the corresponding phenomenal concept. You can't have the concept unless you are acquainted wth the property, and you can't be acquainted with the property unless you have experienced it at first hand.

On the materialist account of phenomenal concepts, by contrast, it comes out as a quite contingent matter that you need an original experience to 'know what it is like'. It does seem that human beings, subject to the Humean qualifications above, can think phenomenally only about experiences they have had before. But the explanation I have suggested for this phenomenon implies that it could well have been otherwise, with different kinds of creatures.

I argued that Mary could not imagine or introspectively classify experiences of red without an original experience because, as I put it, the brain needs an 'original' from which to make a 'mould'. The relevant patterns of neural activation can be fixed initially only by an exogenously caused experience. Now, while this may indeed be true of us humans, it is easy enough to posit creatures who do not work like this, and so are not subject to the same cognitive limitations. These would be creatures who are *born* with imaginative and introspective abilities, so to speak, and do not need any specific experiences to instil them. The necessary 'moulds', and the dispositions to use them, would be 'hard-wired'—that is, would develop independently of any specific experiences. A creature who developed like this would be able to imagine seeing something red, and be poised to classify new experiences as of that type, even before undergoing any exogenously caused red experience. Humans are not like this.[13] But there seems nothing impossible about creatures who are.

2.9 *Imagination and Introspection*

Let me now focus on the relation between imaginative re-creation and introspective classification. So far I have taken it that these two

[13] Perhaps this merits further discussion. As I shall argue in sections 4.4–4.7 below, phenomenal concepts piggyback on perceptual concepts. Humans certainly find some perceptual concepts much easier to acquire from experience than others, and perhaps there are cases at the extreme, perceptual concepts which emerge ontogenetically without any prompting from experience. Current psychological research suggests that this may be true, for example, of the visual concept of a spatio-temporally continuous object (cf. Spelke 1991). If this is right, then perhaps the ability to imagine and introspect visual experiences of objects is independent of experience.

powers go hand in hand, that they are simply two sides of the same coin of 'knowing what it is like'.

Still, on reflection, it does not seem inevitable that the two abilities should accompany each other. There seems nothing incoherent in the idea of creatures in whom they come apart—that is, who can imaginatively re-create experiences, but not introspectively classify them, or who can introspectively classify experiences, yet never imaginatively re-create them. To make these possibilities concrete, simply posit a being with a pattern of neural activation that can be used for introspective classification, yet who lacks anything similar that can be switched on in imagination, or vice versa.

Indeed, we need not go to imaginary creatures to find such dissociations. While it does seem plausible that humans can introspectively classify everything that they can imagine, the link the other way seems far less tight. For example, we seem much better at classifying *smells* introspectively than at re-creating them imaginatively. An actual olfactory experience can create an intense feeling of recognition, yet we may be quite unable to re-create that smell imaginatively later. And to some extent the same point applies across the experiential board. We are rather better at recognizing experiences than re-creating them.

Still, the two kinds of power do tend to accompany each other in humans, which is why it is natural to lump them together under the heading 'knowing what it is like'. Why should this be so, given that it is in principle possible, and to some extent actual, for the two powers to become dissociated? Smells and similar examples show that the correlation between imagination and introspection in humans is by no means perfect. But there remains a question as to why there should be any correlation at all.

An obvious answer is suggested by the models of imaginative re-creation and introspective recognition suggested earlier. Perhaps the same mechanism underlies both powers. That is, perhaps the patterns of neural activity which are 'switched on' in imaginative re-creation are just the same patterns which provide the 'template' for incoming neural signals in introspective classification. So, once some such neural pattern has been fixed by an original experience, it will become available for deployment in *both* imagination and introspection.

This model also suggests a natural explanation of why we aren't always as good at imagining things as we are at introspectively classifying them. We can suppose that, in the standard cases of introspective classification, the 'template' neural pattern will be activated automatically by some initial match with incoming neural signals. In line with the model suggested earlier, the template might then 'send down' further signals, to check whether this initial indication of a match can be 'filled out'.

By contrast, neural activation will not be triggered so directly in the paradigm cases of imaginative re-creation. Rather, acts of imaginative re-creation will result from deliberate choices, or associative connections with other experiences, or from other such sources. If this is right, then we can expect some neural patterns, in some individuals, to be regularly triggered by incoming signals, thus yielding introspective classification, yet to be unavailable to imaginative re-creation. This would happen if there were no links allowing deliberate choices, associated experiences, or anything similar, to excite the relevant neural pattern. Perhaps this is why most of us are no good at imagining smells. We can identify smells all right when we have them, but we have no other way of turning on the relevant pattern of neural activation.

2.10 *Further Issues*

Let me now briefly draw attention to some further questions involving phenomenal concepts. This will enable me to flag some issues to which I shall return later.

2.10.1 *Are Phenomenal Concepts Introspective or Imaginative?*

If introspective classification and imaginative re-creation are separate powers, which can come apart in principle, and to some extent in practice, then ought we to speak of *single* phenomenal concepts of types of experience? For example, I can think about the experience of seeing something red in virtue of introspectively classifying it, and I can think about it by imaginatively re-creating it. So shouldn't we speak here of two distinct phenomenal concepts, an introspective concept and an imaginative concept?

I don't think anything much hangs on this question. It is more a matter of how we describe our data, rather than anything substantial. Accordingly, I shall continue to talk about phenomenal concepts as such, but will take care to distinguish between imaginative and introspective deployments of these concepts when it matters.

2.10.2 *Perceptual Concepts and Phenomenal Concepts*

Later on, in Chapter 4, I shall compare phenomenal concepts with *perceptual* concepts. These are concepts not of experiences of *seeing something red*, or *seeing* an elephant, or *feeling* a circle, but of *redness*, or *elephants*, or *circles*, considered as perceivable features of the non-mental world. I shall argue that there is an intimate relation between phenomenal concepts and perceptual concepts.

Because of this relation, there will be a number of analogies between phenomenal and perceptual concepts. At this stage let me just observe that perceptual concepts, like phenomenal concepts, can be variously deployed both in perceptual *classification* and perceptual *re-creation*. So in this case too there will be some reason to speak of two concepts, a classificatory concept and a re-creative one. But again this will not be a substantial issue, but simply a matter of how we describe our data.

2.10.3 *Theories of Reference*

In section 2.7 I argued that phenomenal concepts are not straightforward indexical constructions, but make essential use of imagination or introspection. But how, then, do they refer? How is it possible for us to refer to conscious experiences by exercises of imagination and introspection? This is a substantial question, which will be crucial for much of what follows. The next chapter, which addresses Saul Kripke's well-known modal argument against materialism, will place constraints on possible answers to this question. (In particular, it will show that phenomenal concepts cannot refer by description.) But it will leave it open how they do refer, and we shall return to this question at length in Chapter 4.

Chapter 3
THE IMPOSSIBILITY OF ZOMBIES

3.1 *Introduction*

Let me take stock of the argument. In the last chapter I argued that we have two quite different ways of referring to conscious properties. We can refer to them using ordinary material concepts, or we can refer to them using distinctive phenomenal concepts which involve special powers of imagination and introspection. I also argued that this conceptual dualism is quite consistent with the ontological monism of Chapter 1. In particular, the causal argument from that chapter is in no way undermined by the existence of phenomenal concepts, since that argument didn't depend on any special assumptions about concepts, but simply appealed to a number of compelling empirical claims, which we have as yet seen no reason to deny.

In short, I am arguing that our thoughts about conscious properties are simply a special case of the situation where two concepts (two 'senses') point to a single referent. Uncontroversial cases of this form are familar enough: the Morning Star = the Evening Star, Cicero = Tully, water = H_2O. Similarly, I say, with conscious properties. For example, we can refer to the experience of seeing red materially, in physiological or psychological terms, or we can refer to it phenomenally, as *this* experience (accompanied by an act of imagination or introspection). But in both cases we are referring to the same real property.

Despite my arguments, I am sure that many readers will remain quite unconvinced. For it certainly doesn't *seem* as if conscious properties are identical to brain properties. Property identity claims involving phenomenal and material concepts are intuitively quite different from ordinary identity claims. There is nothing puzzling about the Morning Star being the Evening Star, or Cicero being Tully, or water being H_2O. By contrast, there is something very counter-intuitive about the phenomenal–material identity claims advocated by materialists. When materialists urge that *seeing red* (and here you must imagine the redness) is identical to some material *brain property*, it strikes many people that this *must* be wrong.

From now on I shall call this natural reaction the 'intuition of mind–brain distinctness'. Materialism needs to say something about this intuition. An intuition on its own may not amount to an argument. But this intuition certainly weighs strongly against materialism with many people. A successful materialism therefore needs to explain this intuition. It needs to show why the conscious mind and the material brain should *seem* so different to us, if they are really the same. Accordingly, I shall address this issue at various points in what follows, and in Chapter 6 shall offer my own preferred explanation of this anti-materialist intuition.

Before that, however, I want to address some more anti-materialist *arguments*. Jackson's knowledge argument is not the only reasoned argument against materialism. Other anti-materialist philosophers have also tried to go beyond the brute *intuition* that mind and brain cannot be identical, and have aimed to show how this denial follows validly from plausible premisses. In this chapter I shall look at Saul Kripke's modal argument against mind–brain identity, and in Chapter 5 I shall look at the argument that materialism leaves us with an unacceptable 'explanatory gap'. (The chapter in between, Chapter 4, will be devoted to further analysis of the structure of phenomenal concepts.)

Neither Kripke's argument nor the explanatory gap argument succeeds in discrediting materialism. But it is worth examining them in detail. A careful analysis of these anti-materialist arguments will add to our understanding of phenomenal concepts.

Moreover, analysing these arguments will help to pinpoint the real source of the intuition of mind–brain distinctness. Most philosophers of consciousness are of the view that this intuition of distinctness owes its currency to one or the other of these anti-materialist arguments. That is, they suppose that some implicit appreciation of these arguments lies behind the widespread feeling that mind and brain must be distinct. I shall show that this diagnosis is mistaken. The arguments are inadequate to explain the intuition, for they apply equally well to cases where we have no intuition of distinctness. In truth, the relation between the intuition and the arguments is the other way round. The intuition stems from a source which is quite independent of the arguments. And then the intuition lends a spurious plausibility to the arguments, since it so strongly predisposes us to believe their conclusion.

3.2 *Epistemology versus Metaphysics*

The initial target of Saul Kripke's modal argument was early post-war materialism, as defended by figures like U. T. Place (1956) and J. J. C. Smart (1959). These early materialists were fond of saying that the identification of mental states with brain processes is a *contingent* identity. By this they meant to convey that the identification rests on empirical evidence, and cannot be established by conceptual analysis alone.

By way of analogy, they invoked scientific identifications like that of *temperature* with *mean kinetic energy*, or *lightning* with *atmospheric electrical discharge*, or *water* with H_2O. Obviously, these identities cannot be established by conceptual analysis alone. A priori reflection on concepts is not going to tell us that temperature is mean kinetic energy. Nevertheless, scientific investigation has shown us that they are indeed the same, and similarly with the other identities. True, the scientific results could have pointed to different conclusions. But they didn't. So it is with mind and brain, said the early materialists. There is no a priori way of showing that they must be identical. But, as a matter of contingent scientific fact, it turns out that they are.

Kripke objected (1971, 1972, 1980) that this doctrine of contingent mind-brain identity is confused. The early materialists were confusing the *epistemological* question of whether mind-brain identities can be established by a priori means alone, or only a posteriori, with the *modal* or *metaphysical* issue of whether the claims thus established are necessary, or only contingent. There is of course nothing wrong with insisting that the relation between mind and brain is an empirical matter, to be assessed in the light of empirical evidence, and not on a priori grounds. But this in itself, insisted Kripke, leaves the modal status of the materialists' claims quite open. There is no legitimate inference from a claim being a posteriori to its being contingent.

Indeed, continued Kripke, once we separate the metaphysics from the epistemology, we can see that the materialists' claim of mind-brain identity would have to be *necessary*, if it were true at all. This is because all identities are necessary. A thing is what it is, and cannot be something else.

It may be a matter of empirical discovery to find out that Cicero is identical with Tully, say, or the Evening Star with the Morning Star, or temperature with mean kinetic energy. But the truths so discovered are necessary truths. To suppose otherwise is to suppose that Cicero might not have been Tully (or the Evening Star might not have been the Morning Star, or temperature might not have been kinetic energy). However, these things make no metaphysical sense. How could Cicero not have been Tully? There is no possible world where Cicero exists, but Tully doesn't. Since they are the same person, Tully will be there if Cicero is. Similarly, you can't have a world with the Evening Star but not the Morning Star, or with temperatures but no mean kinetic energies.

So identities are necessary, if true. In particular, mind-brain identities would have to be necessary, if they were true.[1]

[1] The claim that all true identities are necessary is often qualified by a proviso about rigid designators. I aim to avoid these complications by making the assumption that all genuine identity claims have rigid designators on both sides. Claims like 'John is the tallest man in England' will be assumed not to be identities,

3.3 *The Appearance of Contingency*

So far this mightn't look like much of an objection to materialism. For why can't materialists simply accept Kripke's distinction between epistemology and metaphysics, and agree that mind–brain identities are necessary, while based on empirical evidence? After all, the important point, for the early materialists, was simply that mind–brain identities are a posteriori. Kripke shows that it is wrong to muddle this up with these identities being contingent. So the obvious solution is for materialists to disentangle their metaphysics from their epistemology, and simply agree that their a posteriori identities are necessary.

The trouble now, however, is that these identities don't *seem* necessary at all. Given that Tully = Cicero, a world containing Tully but no Cicero makes no metaphysical sense. If there is only one person, then how *could* he be both present and absent? But there doesn't seem anything similarly incoherent about a world with pains but no brains, or brains but no pains. It seems possible for pains and brains to come apart, in a way that Cicero and Tully simply can't.

Let us suppose, for the sake of the argument, that the materialist wants to identify pains with the firing of nociceptive-specific neurons in the parietal cortex. Then, by analogy with the Cicero–Tully case, it ought to follow that there are no possible worlds with nociceptive-specific neuronal activity but no pains, or pains but no nociceptive-specific neuronal activity. But these things seem manifestly possible. In the actual world, these two states may never come apart. But there doesn't seem anything metaphysically incoherent about creatures who are physically just like us, down to their nociceptive-specific neurons, but who have no feelings of pain. Even less does there seem anything incoherent about a possible

but to have the quantificational form (∃!x) (x is the tallest man in England and x = John). I realize that some philosophers hold that there can be genuine identities involving flaccid designators. But nothing in my arguments depends on this issue, as far as I can see, and it will be far more convenient to assume that all identities are necessary. In line with this, I shall assume for the purposes of this chapter that scientific terms like 'temperature', 'lightning', and 'water' are rigid designators of the first-order kinds or quantities that they pick out in the actual world.

world where there are beings who feel pains, but have no nociceptive-specific neurons.

You might feel that these intuitions of possibility simply reflect the implausibility of identifying pains specifically with nociceptive-specific neuronal activity, rather than with some more abstract or higher-order material property. But this won't wash. It doesn't matter which material property you choose as the candidate for identity with pain (or for identity with whichever other conscious property you may be interested in). It will still seem possible for the conscious feeling and the material property to come apart. To see this, we need only consider the possibility of zombies and ghosts.

Zombies are beings who share *all* our material properties, yet have no consciousness whatsoever. Zombies seem metaphysically coherent, even if never actual. Just imagine a being who is a molecule-for-molecule duplicate of yourself, but who feels nothing at all, who is a mere automaton so far as conscious experience goes. Of course, we don't expect ever to meet such a being. Actual people don't work like that. But still, there seems nothing incoherent about such an insensate doppelganger, who has all your material attributes, yet lacks the conscious ones.

Ghosts are the converse possibility—beings who share *none* of our material properties, yet have just the same conscious states as we do. Again, ghosts seem metaphysically coherent, even if never actual. Just imagine a being who shares your conscious life, yet has no material properties at all, of the kind which underpin conscious life in this world. Such a being would share all your conscious properties, yet have none of your material properties.[2]

If zombies or ghosts are possible, then phenomenal properties cannot be identical with any material ones. Take a generic conscious property C. Possible zombies and possible ghosts both imply that it is possible for C to come apart from M, for any material property you may wish to identify C with. But if this is possible, then it follows that C cannot be identical with any material M. For the dissociation would not be possible if C were really identical with M, any more than it is possible for Cicero to come apart from Tully.

[2] For a discussion of some philosophical asymmetries between zombies and ghosts, see Sturgeon 2000.

Kripke's argument is thus that the *possibility* of conscious properties coming apart from material properties shows that they cannot be identical with material properties. Kripke can of course allow that certain conscious properties are always found hand-in-hand with certain material properties in the actual world. But, from Kripke's point of view, this will mean only that those properties are correlated, not that they are identical. The properties can't be identical, for then there would be no metaphysical sense to the idea that they might come apart—which there clearly is, insists Kripke.

3.4 *Explaining the Appearance of Contingency*

Since materialists are committed to mind–brain identities, and identities are necessary, they need to deny that conscious properties can possibly come apart from material ones. It is difficult, of course, to deny that these things *seem* possible. There doesn't *seem* to be anything metaphysically incoherent about the possibility of zombies and ghosts. But materialists must deny that such things really are possible. So they need to say that zombies and ghosts are a kind of modal illusion. Even though it might seem to us that conscious and material properties can come apart, such dissociations are not really possible.

However, materialists now face another challenge. Why should zombies and ghosts seem possible, if they are not? On the face of it, the mind–brain relation seems quite different from other identities, like Cicero = Tully, precisely in appearing contingent where they do not. Materialists say that this appearance is illusory. But then they surely owe some explanation of this illusion. Since they agree that the mind–brain relation at least *seems* to be contingent, they need to come up with some explanation for *the appearance of contingency*. Why should it seem to us that mind and brain might come apart, when this doesn't seem possible for Cicero and Tully?

There are various ways in which materialists can respond to this challenge. One initially attractive option is to draw an analogy with the scientific identities that the early materialists originally held up as their model of 'contingent identities', like *temperature is mean kinetic energy*, or *lightning is electrical discharge*. We can all agree that, given that these are indeed identities, they can't really be *contingent*, however much they are a posteriori results of scientific investigation. But still, don't they at least *appear* contingent?

At first pass, there certainly seems to be some metaphysical sense to the idea of worlds in which there are temperatures, but no mean kinetic energies, or mean kinetic energies, but no temperatures. What about a world in which sensations of heat turn out to be caused not by mean kinetic energy, but by the flow of some distinct caloric fluid? Isn't this a world in which something other than mean kinetic energy is temperature? Or what about a world in which there are mean kinetic energies all right, but our perceptual apparatus works rather differently, so as to stop us registering mean kinetic energies as sensations of heat? Isn't this a world in which there is mean kinetic energy but no temperature?

Careful readers will realize that, strictly speaking, the last paragraph offers *misdescriptions* of the relevant worlds. Given that temperature actually *is* mean kinetic energy, then there isn't any real possibility of worlds in which temperature and mean kinetic energy come apart. If there is only one quantity here, *it* can't come apart. Strictly speaking, the worlds in which something other than mean kinetic energy causes heat sensations, or in which mean kinetic energy doesn't cause heat sensations, are not worlds in which mean kinetic energy and *temperature* come apart. Rather, they are worlds in which mean kinetic energy (that is, temperature) comes apart from heat sensations. If we feel inclined to describe these as worlds in which mean kinetic energy separates from 'temperature', this can only be because we find it tempting to think of trans-worldly 'temperature' in terms of the symptoms by which we pick out temperature in this world (heat sensations), rather than in terms of temperature's true nature (mean kinetic energy).

Still, this needn't worry those mind–brain materialists who are invoking the analogy with temperature and mean kinetic energy. For their aim is to explain the *appearance* of mind–brain contingency, not its actuality. And the case of temperature and mean kinetic energy would still seem to provide a perfectly good model for this. For, as we have just seen, it is certainly very tempting to speak as if these two quantities can come apart, even if such separation isn't really possible. This temptation may be loose talk, which ties trans-worldly uses of 'temperature' to the symptoms of temperature rather than to its nature. But, even so, it is very natural talk, and it surely suffices to explain the common, if confused, impression that temperatures might not have been mean kinetic energies, or vice versa.

So the idea would be to offer a similar explanation for the apparent contingency of mind–brain identities. Materialists can argue that these identities strike us as contingent only because we are tempted to think of pains in terms of their symptoms rather than their nature. This kind of temptation is why we confusedly think that 'temperature might not have been mean kinetic energy'. Similarly, so the materialist suggestion would go, with the thought that 'pains might not have been nociceptive-specific neuronal activity (or any other material property)'. This thought shouldn't be construed as positing an (impossible) world in which pains separate from their material nature, but simply a world in which that nature comes apart from the symptoms by which we initally pick out pains.

3.5 *Referring via Contingent Properties*

The Kripkean argument isn't finished yet. We need to look more closely at the analogy with temperature and mean kinetic energy. It turns out that it isn't as helpful to the mind–brain materialist as it might seem.

Let us consider more carefully what is going on when we take it that 'mean kinetic energy can come apart from temperature'. To focus the issue, consider a world in which we have different perceptual mechanisms, and so mean kinetic energies fail to cause sensations of relative heat. As I said, the reason why we naturally describe this as a world in which mean kinetic energy is not 'temperature' is that we initially think of temperature as *that quantity, whatever it is, that causes sensations of relative heat*. So, when we specify a world in which mean kinetic energy fails to satisfy that description, it is natural to describe it as one in which mean kinetic energy is not 'temperature'. This is a misdescription, given that 'temperature' actually *is* mean kinetic energy—such a world is rightly described as one in which mean kinetic energies do not 'cause heat sensations'. But, as we saw, it is a very natural misdescription.

On the standard contemporary view of such scientific identities, the pre-theoretical terms involved refer by description.[3] Thus, everyday terms, like 'water' or 'temperature' or 'lightning', will pick out *that* quantity or property *which* satisfies some everyday description. Prior to scientific investigation, we won't yet know which property this is—that is, we won't yet know that the relevant term names H_2O or mean kinetic energy or electrical discharge. These things are for science to discover. So initially the reference of the everyday terms will be fixed via some pre-theoretical description, like 'odourless, colourless, and tasteless liquid', 'causing heat sensations', or 'flashing through the sky before thunder'. These descriptions will be associated a priori with our initial terms.

However, the properties which these descriptions invoke (causing heat sensations, and so on) will be possessed only contingently by the referents. *Temperature* (that is, mean kinetic energy) has the property of causing heat sensations contingently. This shows itself in the fact that there are genuinely possible worlds in which mean kinetic energies (that is, temperatures) do not cause heat sensations, because of alterations in our sense-organs. These are the worlds which we are tempted to describe inaccurately as having 'mean kinetic energy but not temperature', though strictly they are only worlds where 'mean kinetic energy fails to cause heat sensations'.

The important point in all this is that we get an appearance of contingency with scientifically established identities only because the everyday terms involved in such identities have their references fixed by contingent properties. Because these properties are *contingent*, there are genuinely possible worlds in which the scientific

[3] When I talk about 'reference by description', as I frequently shall from now on, I should be understood in a generous sense. I make no assumption that the relevant term's reference is, or was, fixed by explicit stipulation, or even that the description can be articulated in a public language. I shall mean only that the term is a priori equivalent to something of the form 'the (actual) x which is Ø', where 'Ø' represents some independently referring concept.

referents lack these properties, such as, for example, worlds in which mean kinetic energies do not cause heat sensations. Because these properties *fix reference*, it is natural to describe these as worlds in which the scientific referents come apart from the everyday terms, as worlds in which mean kinetic energy is 'not temperature', even though, strictly speaking, this is an inaccurate description.[4]

This suggests that if materialists are to run the same story with mind–brain identities, they will have to hold that pre-theoretical terms for conscious states, like 'pain', pick out their referents via contingent properties. This is where Kripke's argument really bites. For, when we try to run this model, it turns out not to work.

On this model, 'pain' will have to fix its reference to some material property M via some contingent feature of that property. The idea would have to be something like this: the material property M, the real referent of 'pain', is picked out as that property, whichever it is, that contingently generates painful reactions in humans. Given this, there would then seem to be space for a world in which *pain* (that is, M) does not generate those painful reactions—for example, a zombie world in which there are beings who share our Ms but not our painful reactions.

And the explanation for the apparent contingency of mind–brain identities would then need to run as follows. 'It may be natural to describe the zombie world as one in which there are "Ms but not pains". Indeed, this explains our impression that the relation between pain and M is contingent. But this relation is not really

[4] As it happens, I myself am not particularly persuaded that pre-theoretical kind terms like 'water' have their references fixed by description. (For a strongly opposed view, see Millikan 2000.) On the other hand, it is a moot point whether this assumption about reference fixing is essential to the suggested explanation of the apparent contingency of water = H_2O. The crucial requirement is that 'water''s associations with contingent descriptions can lead us to confuse possibilities where those descriptions are/aren't satisfied with possibilities where water is/isn't present; whether these associated descriptions also fix reference in the actual world seems a further question. Still, in the light of all this, it will do no harm to go along with the standard Kripkean literature in adopting the reference-fixing model, given that the rest of the Kripkean story will go through even if the model is wrong. I would like to thank Scott Sturgeon for help on this point.

contingent, for we are misdescribing the relevant world: the zombie world is not one which lacks *pains*—it just lacks the further contingent property by which we pick out pains in this world: namely, the generation of painful reactions.'

But now something seems to have gone wrong. For surely the zombie world lacks *pains*, not just 'painful reactions'. For what are pains, except the 'painful reactions generated in humans'? The 'generation of painful reactions' can't plausibly be viewed as some contingent property which helps us to pick out pains in the actual world. Surely it is the essence of pain itself.

Stubborn materialists may feel inclined to dig in their heels here, and insist that zombies *do* have pains. That is, they could insist that pain itself *is* different from 'painful reactions'. Pain is identical with some material state, and so is present in zombies. It just fails, in the zombie world, to generate those subjective 'painful reactions' with which pain is contingently associated in the actual world, and which we happen to use to fix the reference of our word 'pain'.

But this ploy not only requires a quite implausible account of the working of the concept 'pain'—it doesn't help anyway. For anti-materialists can now simply switch their attack to the 'painful reactions' themselves. It is agreed on all sides, and in particular by the materialists, that zombies lack *these*. Yet having painful reactions is itself a conscious property, present in humans who are in pain in the actual world. Anti-materialists can simply point out that this in itself refutes materialism. If all conscious properties are material, then how come zombies, who are stipulated to share all the material properties of humans, so much as lack 'painful reactions'? Materialists need to identify 'painful reactions' themselves with some material property, and so can't coherently suppose that zombies lack them.

So, whichever way they turn it, materialists seem unable to offer a satisfactory account of the apparent contingency of mind-brain relations. It looks as if they have no option but to admit that these relations really are contingent: even if properties like pain are perfectly correlated with certain material properties in this world, this correlation could fail to obtain in other possible worlds. But then it follows that conscious properties aren't *identical* with material properties. For identities, unlike correlations, cannot come apart in other possible worlds.[5]

3.6 *A Different Explanation*

Materialism may be down, but it is by no means out. There is another way to respond to Kripke's challenge.

There is nothing wrong with most of Kripke's argument. Identities are indeed necessary. So materialists must deny that it is possible for conscious properties to come apart from the material properties they are identical with. At the same time, it certainly *seems* possible that these properties should come apart. So materialists owe an explanation of this appearance of contingency. Yet it won't do to say that phenomenal concepts like 'pain' pick out their referents via

[5] There might seem to be another way for materialists to appeal to the idea of reference by contingent description. Why not apply this idea to the *material* concepts involved in mind–brain identities, rather than to the phenomenal terms? After all, it is quite natural to hold that such concepts refer to physical kinds *as entities which play specified theoretical roles*. So can't the materialist construe a zombie world as one where those theoretical roles are filled by something other than the stuff which fills them in the actual world? *Pain* could then be identified with something involving that actual stuff, and zombies wouldn't feel pain because they lacked that actual stuff, while satisfying the contingent descriptions which pick it out in this world. Well, I am happy enough to accept that such creatures are possible—that is, beings who lack pains because they lack the actual stuff picked out by scientific descriptions in this world. But I don't think this helps materialists to explain the apparent possibility of *zombies*. For, by my lights, these creatures aren't proper zombies, but only pseudo-zombies. This is because I take the stuff picked out by scientific descriptions in the actual world to be physical stuff (cf. Ch. 1 n. 5). So proper zombies, unlike pseudo-zombies, will share this physical stuff, and so can't lack pains, if materialism is true. In order for pseudo-zombies to help explain why it *seems* possible for real zombies to lack pains, materialists will need to argue that this illusion stems from a confusion of pseudo-zombies, who do lack pains, with real zombies. But it is surely implausible that there would be no intuition of distinctness without this confusion. Fix your mind clearly on real zombies, who do have the physical stuff picked out by scientific descriptions in the actual world. Doesn't the intuition of distinctness still persist for such beings, even when you aren't confusing them with pseudo-zombies? That is, don't real zombies, who have the same physical stuff as us but lack pains, seem just as possible as pseudo-zombies?

contingent descriptions, à la 'temperature'. This claim itself turns out to be inconsistent with materialism.

The loophole is that the contingent description story isn't the only way to account for the appearance of contingency. The materialist can agree with all the Kripkean points listed in the last paragraph, yet offer a different explanation for the appearance of contingency, one which doesn't have phenomenal concepts referring by contingent description.

Before explaining how this might work, it is worth emphasizing exactly why the contingent description story is a poisoned chalice for materialism. At its most graphic, the challenge facing materialism is to explain why zombies seem not to satisfy the concept 'pain', even though, according to materialism, they must. Anti-materialists face no problem here—they simply hold that 'pain' refers to some non-material property which zombies lack (so zombies don't just *seem* not to satisfy 'pain'—they don't). Obviously, materialists need a different story. However, as soon as they accept Kripke's invitation to assimilate the concept 'pain' to concepts that refer by contingent description, they are in trouble.

For the only descriptions that can fill the bill here are descriptions associating 'pain' with further phenomenal concepts ('painful reactions'). After all, we have agreed, following Jackson's argument, that 'pain' itself is a *phenomenal* concept, and as such is distinct from any material concepts. So there is no question of 'pain' referring by association with material concepts (say, to the state that produces certain behavioural reactions); this would simply make 'pain' a material concept itself. If 'pain' refers by description, the descriptions must involve phenomenal concepts, not material ones. However, now materialism is stuck. For the Kripkean analysis now invites them to say that zombies seem to have no pains only because they do not satisfy these associated phenomenal concepts. And this takes them back to where they started. For how can zombies lack these further phenomenal features, despite their physical identity with feeling humans? Here we seem to be left with no alternative to the anti-materialist claim that humans have extra phenomenal properties that the physically identical zombies lack.

Materialists should refuse Kripke's invitation. They should say that phenomenal concepts refer *directly*, and not by description.[6] Phenomenal concepts don't pick out their referents by invoking certain further features of those referents, but in their own right, so to speak.[7]

This claim of course raises questions about *how* phenomenal concepts do this. How do phenomenal concepts pick out their referents, if not by description? But let us not pause to answer this question at this point. It involves a number of issues, which I shall discuss at length in the next chapter. For now it is enough to note that materialists must somehow assume that phenomenal concepts refer directly, if they are to avoid Kripke's trap.

Given this, materialists can aim to construct a different explanation for the appearance of mind–brain contingency. To start with, they can point out that we have these two quite different ways of referring to conscious properties. We can refer to them with phenomenal concepts or with material concepts. And this in itself, materialists will maintain, can generate the impression that phenomenal properties might be different from material properties.

After all, materialists can point out, the distinctness of phenomenal and material concepts certainly makes it *conceivable* that zombies and ghosts should exist. Since there are no a priori connections between phenomenal and material concepts, there is no conceptual contradiction in positing beings with all relevant material properties but no conscious ones, or vice versa. Materialists must of course deny that the conceivability of these things shows that they are really possible. But they can still maintain that it shows why they *seem* possible. Zombies and ghosts seem possible, materialists can thus say, simply because they can be imagined

[6] When I talk about 'direct reference', as I frequently shall from now on, I will simply mean reference that is not by description.

[7] True, there is the possibility that some phenomenal concepts could refer by descriptions involving *other* phenomenal concepts: 'pain' could refer to that property *that generates feelings of painfulness*. But, apart from the fact that this doesn't avoid Kripke's trap, as we have just seen, this can't be true of all phenomenal concepts, on pain of regress. Accordingly, I shall ignore this buck-passing possibility in what follows. The interesting philosophical issues relate to those primitive phenomenal concepts that do refer directly, not to any derivative ones that then piggyback by description.

without violating any a priori, conceptual constraints, and not for any other reason.

3.7 *Thinking Impossible Things*

This might seem a bit quick, and indeed it is. Let me go a bit more slowly. There are a number of further issues raised by this materialist response to Kripke.

For a start, some may want to question whether the response really makes sense. Suppose materialists are asked to explain *what* people are thinking, when they entertain the possibility of zombies and ghosts? What possible worlds provide a content for these thoughts?

In general, when we think something, the content of our thought can be equated with some set of possible worlds: namely, those possible worlds whose actuality would make the thought true. On this model, a true thought is one whose content includes the actual world, while a false thought is one whose content does not. And necessary thoughts have all worlds as their content, since they are true in any world. However, this account of content leaves no room for thoughts about genuine *impossibilities*, since there are no possible worlds whose actuality would make an 'impossible thought' true.

True, we have seen that there is one sense in which people can sometimes be held to be thinking about impossibilities. Take thoughts like 'mean kinetic energy might not have been temperature'. Now, this description represents an impossibility. Still, it can be regarded as a *mis*description of what is really being thought. That is, we can re-describe the thought as really answering to a genuinely possible world: namely, a world in which mean kinetic energy does not cause heat sensations. Similarly with 'water might not have been H_2O', or 'lightning might not have been electrical discharge'. In each case, we can 'reconstrue' these strictly 'impossible thoughts' as laying claim to genuine possibilities: namely, possibilities in which the relevant properties fail to satisfy the descriptions which everyday terminology uses to pick them out.

But note now how this story requires that the relevant properties are picked out by descriptions to start with. Without these descriptions, there wouldn't be any real possibilities to breathe content into the 'impossible thoughts'. So this now gives us a new version of the Kripkean challenge. If phenomenal concepts pick out their material referents directly and without description, as the materialist now has it, then how can we so much as think that consciousness might come apart from material properties? For, on the current materialist story, we would not only be thinking a strict impossibility, but there would be no descriptions around to transpose this impossibility into a genuinely thinkable possibility.

In response to this challenge, materialists should simply deny that there is any difficulty about thinking genuine impossibilities. Sometimes our thoughts answer to no possibility. This doesn't stop them being thinkable, even when there is no other possibility around to give them an alternative content. (Alice laughed. 'There's no use trying,' she said: 'One *can't* believe impossible things.' 'I dare say you haven't had much practice,' said the Queen.)

To see how it might be possible to think impossible things, consider a case involving proper names. Suppose Jane is familiar with the names 'Tully' and 'Cicero', but doesn't know that they name the same person. Suppose, moreover, that Jane has no specific beliefs involving these terms. She has picked up the names from other people, and to this extent is competent to use them, but beyond that has no special knowledge of Cicero or Tully. Now Jane may well entertain the thought that Cicero is not Tully. Indeed she may well *believe* that Cicero is not Tully. But note that she will be thinking an impossible thought here. There is no possible world corresponding to her thought, no world in which Cicero is not Tully.

Even so, Jane can surely think this thought. And she can do so even though she doesn't associate any descriptions with 'Cicero' or 'Tully', and so can't be thinking about some other possible world, some world in which Cicero/Tully doesn't satisfy those descriptions. I take this example to show that there can be thoughts with impossible contents. Indeed, it gives us a simple recipe for constructing them. Take two names for the same thing, join them in a thought where they flank a term for non-identity—and there you are.

I say the same thing about denials of mind-brain identity. Here we have names for properties, rather than people. But the point is just the same. Two terms—a phenomenal term and a material term—can name the same phenomenal/material property. There is thus no real possibility of non-identity. But this doesn't stop us forming contentful thoughts about non-identity, such as that pain is different from any material property you care to choose. Nothing further is needed to explain the existence of such thoughts. Just take phenomenal concepts, material concepts, and a term for non-identity—and there you are.

Indeed, having come this far, we can see that we may as well have said the same thing about imagining such impossibilities as that temperature is not mean kinetic energy. There is no real need to tell the complicated Kripkean story about our really imagining something else—namely, a world in which mean kinetic energy/temperature lacks the properties which fix reference to that quantity in this world. Why not simply say that 'temperature' and 'mean kinetic energy' are different concepts, which they clearly are for most people, and use this fact alone to explain how people can think the impossible thought that temperature is not mean kinetic energy without any conceptual inconsistency. The point, once more, is that there is nothing difficult about thinking an impossible thought, once you have two terms for one thing.

Of course, there remains a genuine disanalogy between cases like temperature-mean kinetic energy and the mind-brain cases. Since 'temperature' arguably refers via a description, there is indeed a genuine further possibility in the offing—that mean kinetic energy/temperature does not satisfy that description—even if we don't *need* this possibility to provide a content for thoughts that temperature is not mean kinetic energy. By contrast, there is no genuine further possibility corresponding to the thought that zombies might have no feelings. Since phenomenal concepts don't refer by description, there is simply no genuine possibility associated with the thought that a being may share your physical properties yet lack your conscious ones.

3.8 *Conceivability and Possibility*

Some philosophers hold that, contrary to the claims just made, conceivability always guarantees a real possibility. They maintain that, to every conceivable non-identity ($N \neq M$, say), there corresponds a genuine possibility. In cases where N *is* M, this can't of course be the possibility that N is not itself. Rather, in such cases, it must be that N (or M) refers by association with contingent descriptions, which then generates the possibility that the entity referred to might not satisfy those descriptions.

Putting all this together, these philosophers thus hold that, whenever $N \neq M$ is conceivable, either (a) one of the terms involved refers by description, or (b) N really isn't identical to M. This makes it clear why materialists must deny the initial premiss: they cannot allow that conceivability always points to a real possibility. For, if it did, then the manifest conceivability of zombies would imply either (a) that phenomenal concepts refer by contingent description, or (b) that phenomenal properties aren't material properties. But the former alternative is ruled out by Kripke's argument, and the latter refutes materialism straight off.[8]

My response is that conceivability does not always point to a real possibility. I take the Cicero–Tully example from the last section to provide strong support for this view. It is conceivable, for Jane, that Cicero $\neq$ Tully, even though (a) Cicero *is* Tully and (b) she associates neither Cicero nor Tully with any descriptions.

Someone who wants to uphold conceivability as a guarantee of possibility will need to argue here that Jane must have some further ideas about Cicero and Tully, if she is to have genuine concepts of them. That is, she must associate certain descriptions a priori with 'Cicero' and 'Tully', if she is really to be capable of thinking with these terms. This will then restore the link between conceivability and possibility, since it will give us the possibility that Cicero/Tully does not satisfy those descriptions.

But why suppose that any such associations are necessary for Jane to be competent with these terms? The theory of names is a large

[8] For completeness, we should also consider the option that material concepts might refer by description. But we saw in n. 5 that this goes nowhere.

subject, and this is not the place to start pursuing it. But one clear lesson of the last thirty years of work in this area is surely that Jane's conceptual competence with 'Cicero' and 'Tully' need owe nothing to any specific ideas she associates with these terms. Rather, it will be enough if she has picked up the names 'Cicero' and 'Tully' from competent speakers, and intends to use them as they do. And this clearly doesn't require that she associate any further descriptions with these names.

More generally, the contention that conceivability is a guide to possibility places implausibly strong constraints on the theory of reference. It requires that, whenever two directly referring terms refer to the same thing, it must be a priori knowable that they do so. For, on the conceivability → possibility assumption, if it is so much as *conceivable* that some *directly* referring 'N' and 'M' do not co-refer, then it must be *true* that $N \neq M$, for without any associated descriptions there is no other possibility around to explain the conceivability. On the conceivability → possibility view, then, we can be confident that two entities really are distinct whenever directly referring thoughts about them allow them to *seem* possibly distinct.

From now on I shall use the term 'the transparency thesis' for the claim that identities involving two directly referring terms are always a priori knowable. I see no reason whatsoever to accept this thesis. It seems to me to hinge on some atavistic view of reference. For the transparency thesis to be true, the basic referential relations, direct referential relations, would have to involve some kind of unmediated mental grasp of the entities referred to, a grasp which left no room for mistakes about identity.[9] Far from accepting this, I take the basic referential relations to depend on all kinds of facts external to thinkers' heads, facts which create plenty of room for a thinker to be wrong about whether two terms refer directly to the same thing.

Let me conclude this section with a historical observation. There is something ironic in the fact that the works in which Kripke first elaborated the anti-materialist modal argument are also the works in which he first defended the causal view of proper names. For the

[9] For an extended critique of this view of direct reference, see Millikan 1993, 2000.

modal argument is only compelling, as we have just seen, as long as there are no impossible thoughts involving only directly referring concepts. Yet, if Kripke's causal view of names is right, proper names provide the most obvious counter-example to this thesis. I don't know what to make of this curiosity. Perhaps the moral is that it takes time for things to become clear in philosophy, even to the most penetrating minds.

3.9 *The Intuition of Distinctness*

Let me finish this chapter by returning to the 'intuition of mind–brain distinctness'. So far I have offered an explanation of how zombies and ghosts can seem possible, even though they are not. Some philosophers think that this story also explains why mind–brain distinctness should seem intuitively compelling, even though it is false. (Cf. Hill 1997, Hill and McLaughlin 1998.) But this is a mistake. The intuition of mind–brain distinctness has an independent source, quite separate from the modal issues to which Kripke's argument draws attention.

As a preliminary to showing this, note that there is a significant disanalogy between the Cicero≠Tully case and the mind≠brain case. When I introduced Jane as an example of someone who could think an impossible thought, I took care to make her *ignorant* of Cicero's identity with Tully. She was capable of thinking the impossible non-identity precisely because she had no reason to think Cicero and Tully the same person. By contrast, someone who *does* accept that Cicero is Tully will cease to think that there is any possibility of distinctness, as I pointed out right at the beginning of this chapter. This person will no longer be able to make any good sense of the possibility that Cicero might exist, but not Tully. Since they are the same person, Tully will be there if Cicero is.

This point does not undermine the use I have made of the Cicero≠Tully example. Even if the non-identity appears possible only while Jane remains ignorant, the ignorant Jane still gives us an example of someone who can think an impossible thought, a thought to which no genuine possibility corresponds.

Still, the ignorance-dependence of Jane's ability to think her impossible thought does mark a contrast with the mind-brain case. For, in the mind-brain case, the impression that mind and brain might come apart is likely to persist, *even* among those who profess the view that they must be identical. Take my own case. I would say that I am persuaded, by the arguments you are reading in this book, that mind and brain must be identical. Yet zombies and ghosts still strike me as being intuitively possible. I don't seem to have any trouble grasping zombie or ghost scenarios. Isn't this just the idea of phenomenal states without material states, or vice versa?

Given the analysis in this chapter so far, this disanalogy should appear very puzzling. Now that I know that Cicero = Tully, I can no longer make any good sense of the suggestion that Cicero might not have been Tully. What am I supposed to imagine? That he might not have been himself? But if I can't make sense of this possibility, then I ought not to be able to make sense of zombie and ghost possibilities either. If I accept that pain is identical to some material state, as I say I do, then oughtn't I to find zombies and ghosts as incoherent as Cicero without Tully? After all, what am I supposed to be imagining? That pain might not have been itself? My analysis so far may have explained why this 'impossible thought' will occur to people who do *not* accept mind-brain identity. But it leaves us with a puzzle about why it should persist in people, like myself, who say that they do.

However, I don't think that this puzzle shows that there is anything wrong with my analysis of the Kripkean argument. Rather, it shows that somethings stops us *really* believing the materialist identification of mind with brain, even those of us who officially profess materialism. And this is why even we materialists continue to feel that zombies are possible. We aren't fully convinced that phenomenal properties are identical with material ones, and to this extent naturally continue to think that they can come apart, as in zombies and ghosts. This is where the mind-brain case is different from the Cicero-Tully case. There is no corresponding barrier to fully accepting that Cicero = Tully. So plenty of people do fully accept this identity, and as a result cease to be able to make any sense of Cicero and Tully coming apart.

Why should it be so hard fully to embrace mind-brain identity? In my view the obstacle is a basic anti-materialist intuition, whose source is quite independent of the issues considered in this chapter. This intuition continues to operate even in those, like myself, who are otherwise persuaded that there are good arguments for materialism, and stops us really believing the materialist conclusion. This is the intuition of mind–brain distinctness which I mentioned at the beginning of this chapter. As I said, I shall offer my own account of why this intuition should be so compelling, even though false, in Chapter 6. At this stage I want only to emphasize that the intuition of distinctness in no way derives from the complex modal considerations discussed in this chapter. It is simply a direct intuition that phenomenal properties are different from material properties. It is this direct intuition that makes it hard for us to identify phenomenal properties with material properties, even given the causal argument for materialism, by contrast to the way that we readily accept that Cicero = Tully, once we are shown evidence for that identity.

So I deny that we are so intuitively persuaded of mind–brain distinctness because we are somehow moved by Kripke's argument. And, on reflection, this account of the source of the intuition is surely quite implausible. The idea would have to go something like this: first we feel the pull of an initial modal judgement, that zombies and ghosts are possible; then we note that this modal judgement can't be dismissed as an illusion along the lines of the temperature and mean kinetic energy case; from this we infer that zombies and ghosts really are possible; and thence we conclude that phenomenal and material properties must be different, notwithstanding any causal reasons for identifying them.

For my money, this is clearly too high-falutin' to explain the widespread conviction that the conscious mind must be separate from the brain. (Moreover, if this were the right story, then we ought equally to end up convinced that Cicero and Tully cannot actually be identical, despite contrary evidence, since here too any initial modal judgement of possible distinctness will resist explanation by analogy with temperature and mean kinetic energy.) So in due course I shall offer a quite different explanation for the intuition of distinctness, an explanation which owes nothing to Kripke's modal analysis.

Chapter 4
PHENOMENAL CONCEPTS

4.1 *Introduction*

I have argued that materialists should be conceptual dualists. At the ontological level, of course, they must be monists. But at the level of concepts they should distinguish two different modes of reference to phenomenal/material properties. In addition to the possibility of referring to these properties *as* material, they should also recognize a distinct mode of referring to those properties via phenomenal concepts which pick them out, so to speak, in terms of the way they feel.

The initial reason for recognizing phenomenal concepts was Jackson's knowledge argument. Despite its original intention, this argument failed to demonstrate that phenomenal properties are non-material. But it did establish the existence of non-material *concepts*, ways of referring to conscious experiences which are standardly available to human beings only after they have actually undergone the experiences in question.

Kripke's modal argument then told us something further about phenomenal concepts. They must somehow refer directly, not via description. They don't identify their referents as the bearers of some further property that they may contingently possess.

So far, then, we know two things about phenomenal concepts. Their possession is standardly consequent upon some earlier version of the type of experience they refer to. And they refer to that experience directly, and not via some description. In this chapter I want to build on this basis, to develop a more detailed understanding of the structure and referential power of phenomenal concepts.

4.2 *Psychological, Phenomenal, and Everyday Concepts*

It will be helpful to start by clarifying the relationship between everyday thought and the conceptual dualism that I have been urging. I want to distinguish between phenomenal and material concepts of experiences. However, this distinction plays no prominent role in everyday thought. Everyday discourse uses undifferentiated words for phenomenal states, like 'pain' or 'hearing middle C', and does not stop to specify whether these words should be understood as expressing phenomenal concepts or material ones.

I think we should view everyday words like 'pain' and 'hearing middle C' as simultaneously expressing both sorts of concepts. Before considering exactly how this expressional duality might work, let me be a bit more specific about the kinds of *material* concepts that might plausibly be expressed, along with phenomenal concepts, by everyday discourse. The relevant concepts here will standardly be concepts associated with causal roles, concepts that pick out their referents in terms of a structure of macroscopic causes and effects (such as bodily damage and avoidance behaviour in the case of pain, or ambient sounds and musical responses in the case of hearing middle C). While experts like Mary may often know a lot more about the physical *realization* of such causal structures in specific kinds of beings, I take it that such detailed physical information is not normally part of pre-theoretical, everyday thought. So, in so far as everyday thought utilizes non-phenomenal material concepts of experiential properties, these will be concepts which pick out their referents in terms of everyday causal roles, not concepts involving specific physical constitutions. Following David Chalmers (1996), I shall call these 'psychological concepts'.[1]

So the thought I wish to pursue is that an everyday term like 'pain' expresses *both* a phenomenal concept of pain, a concept of a state that feels a certain way, so to speak, *and* a psychological concept of pain, a concept that refers by association with a certain causal role. Does this mean that the everyday term 'pain' is equivocal, expressing two quite different ideas, which careful users of the language need always to disambiguate? Well, there is a kind of ambiguity in play here. But it is importantly different from the paradigm case of ambiguity, where a given syntactic form ('bank' or 'bat') expresses two quite unrelated concepts which refer to two quite different entities. For the presumption of everyday thought is surely that the two concepts, the phenomenal and the psychological concepts of pain, are connected by the fact that they actually refer to the same state.[2]

To see how this might work, consider the case of multi-criterial concepts, of the kind often found in science, where two independent criteria (*resistance to acceleration* and *gravitational charge*, say) are both regarded as diagnostic of some kind (*mass*). Now, such multi-criterial concepts display a certain kind of ambiguity, in that in many such cases it will be left indeterminate how exactly the different criteria fix the referent. For a body to have a given mass *m*, must that

[1] As we have seen before (Ch. 1 n. 11, Ch. 2 n. 2), role concepts can be of two kinds: they can name whichever property realizes the role, or they can name the higher property which constitutes the role. I shall leave it open which kind psychological concepts fall under.

[2] Scott Sturgeon has pointed out to me that there are some further possibilities here, because of role-realizer complexities. Perhaps everyday thought does not assume that phenomenal and psychological properties are strictly identical, but rather that the one kind realizes the other. This would account equally well for the undisambiguated everyday usage of experience terms like 'pain'. The most obvious option here would be that psychological concepts are taken to name roles, and phenomenal concepts are taken to name first-order properties which realize those roles. But nothing I have said rules out the converse possibility either, that *phenomenal* concepts are taken to name roles, while psychological concepts are taken to fix reference to first-order states which realize those roles. Still, nothing will be lost for present purposes if we skip over these complexities, and simply continue to credit everyday thought with the general assumption that phenomenal and psychological concepts 'co-refer'.

number measure (a) the body's resistance to acceleration, or (b) its gravitational charge? Or perhaps the requirement should be that *m* measures (c) *both* the body's resistance to acceleration *and* its gravitational charge. Or, again, perhaps it should measure (d) *either* the body's resistance to acceleration *or* its gravitational charge.

These are all quite distinct suggestions about the precise meaning of 'mass', as is shown by the fact that there are possible worlds in which these different suggestions fix different extensions. (Consider a possible world in which some body has resistance to acceleration *m* but a different gravitational charge. Then it has mass *m* according to (a) and (d), but not according to (b) or (c). Again, consider a possible body with gravitational charge *m* but a different resistance to acceleration. Then it has mass *m* according to (b) and (d), but not according to (a) or (c).)

Note, however, that there was no real pressure for Newtonian physicists to decide between these different options. This was because they believed that gravitational mass and resistance to acceleration are always equal in the actual world, and so were confident that options (a)–(d) would always give the same answers for any actual bodies.

This is a typical situation in science. The different criteria associated with multi-criterial concepts each pick out the same quantity; in consequence, scientists who take the criteria so to work in concert will see no need to do any semantic refining. Newtonian physicists never felt obliged to decide between resistance to acceleration and gravitational charge as criteria for mass, precisely because they believed that the same quantity would be picked out either way. Note how this kind of benign ambiguity is different from the 'bank' or 'bat' kind of case, where some term clearly has quite different referents on alternative readings. By contrast, in the typical scientific case the same entity is picked out on all the possible readings, which is why the ambiguity is of no concern to the users of the term. (See Papineau 1996 for a detailed discussion of this issue.)

Similarly, I say, with everyday discourse and 'pain'. The term 'pain' does indeed express two conceptually independent notions, phenomenal and psychological. But since it is generally assumed that these two concepts refer to the same property, everyday thought does not exert itself to decide which concept the term 'pain' *really* expresses.

Of course, semantic refinement of a multi-criterial concept can become mandatory if new discoveries overturn the empirical assumption that the different criteria all pick out the same kind. When general relativity showed scientists that resistance to acceleration and gravitational charge can come apart, it was no longer possible to work with the old unrefined Newtonian concept of *mass*, and scientists were forced to distinguish inertial mass from rest mass.

Similarly, theoretical developments in psychology could force everyday thought to refine its undiscriminating usage of terms like 'pain'. If such developments showed that the phenomenal and psychological concepts expressed by some everyday conscious concept fail to co-refer, then everyday usage would need somehow to refine its terminology.

As we have seen, it is an entirely a posteriori matter whether the phenomenal and psychological concepts associated with everyday phrases like 'pain' or 'hearing middle C' refer to the same property. Issues of such co-reference answer to empirical evidence. In Chapter 7 I shall look at the kind of empirical research that can decide such questions. We shall see there that certain complications arise with empirical research into phenomenal properties: phenomenal concepts turn out to be vague in certain dimensions, and this prevents precise answers to some questions about their material referents. But this vagueness arises mainly when we try to stretch our concepts beyond normal human beings to differently constituted creatures. In connection with normal humans, there is no reason why the empirical evidence should not show associated phenomenal and psychological concepts to coincide referentially. Not that we should take it for granted that this will be the result. It may be that everyday thinking is mistaken in various respects about how the referents of phenomenal and psychological concepts line up with each other,[3]

[3] Thus consider the case of 'morphine pain' (Dennett 1978*b*). Patients who take morphine to allay some pre-existing pain will characteristically report that they can still feel the pain—it hasn't gone away— it just doesn't disturb them any more. Now, the interpretation of such remarks is not uncontroversial. But one natural

and in such cases the results of empirical research will require revisions in everyday usage, just as post-Einsteinian empirical research required physicists to revise their use of 'mass'. But there is no reason to expect this to be the norm, and in most cases we can anticipate that everyday usage is empirically entitled in its suppositions that associated phenomenal and psychological concepts pick out the same properties in normal humans.

Empirical research takes place in a context of general metaphysical presuppositions. In particular, the inferences you draw from observed correlations between applications of phenomenal and psychological concepts will depend on your general metaphysical attitude to the relation between mind and brain. In this connection, it is interesting to observe that, while both materialism and dualist interactionism will support conclusions about phenomenal–psychological co-reference, epiphenomenalism will not. Even given ideal correlations between applications of phenomenal and psychological concepts, epiphenomenalists will deny phenomenal–psychological co-reference. It follows that, while materialism and interactionist dualism promise to uphold existing usage, as I have analysed it, epiphenomenalism will require wholesale reformation of that usage.

The point is that both materialism and interactionist dualism allow that a phenomenal and a psychological concept can pick out just the same property. Consider the word 'pain', for example. As I have it, this expresses both a phenomenal concept (how *pain* feels) and a psychological one (involving causal mediation between damage and avoidance). Both materialism and interactionist

way of reading them is as showing that the phenomenal and psychological concepts of pain are not coextensive. The phenomenal concept is here satisfied—the state *feels* like pain—even though the psychological concept is not—the state doesn't agitate the subjects or generate desires for its cessation. Again, consider Benjamin Libet's experiments on decisions to act, as described in section 1.4 above. Again, the precise interpretation of this experiment is open to debate, but once more a natural interpretation is that it discredits the assumption that phenomenally identified decisions always coincide with psychologically identified ones. Here the event that *feels* like a decision, as subjectively identified, doesn't satisfy the psychological notion of a decision, for a central strand in the psychological notion is surely that decisions causally *initiate* actions, and the phenomenally identified decision is too late for that.

dualism will take the phenomenal concept here to refer to a property which does so causally mediate. True, where materialism will hold this to be a material property, interactionist dualism will take it to be some non-material property with the power to produce physical results. But, for all that, they will agree that the phenomenal concept associated with 'pain' identifies the causally potent property that is picked out by the psychological concept (or to a property which realizes that property—cf. n. 1 above).

But not so according to epiphenomenalism. For, on the epiphenomenalist view, the referent of the phenomenal concept associated with the everyday term 'pain' is an inefficacious phenomenal property. And this is distinct from the material property which mediates causally between damage and avoidance, and so satisfies the psychological concept associated with 'pain'. So, on my analysis, epiphenomenalists face a terminological decision. One option would be to decide that 'pain' (a) refers to the inefficacious epiphenomenal property, which presumably satisfies the phenomenal concept, but not the psychological causal role concept. Another would be to hold that 'pain' (b) refers to the efficacious physical property, which satisfies the psychological causal role concept but not the phenomenal concept. Again, there is the option of saying that (c) nothing satisfies the word 'pain', on the grounds that nothing fits *both* the phenomenal concept *and* the psychological concept of pain. Or perhaps 'pain' should be taken to (d) refer to anything that satisfies *either* the phenomenal concept *or* the psychological concept of pain.

Perhaps the first of these options strikes you as most plausible—epiphenomenalists should say that 'pain' refers to their putative inefficacious epiphenomenal properties. But, in any case, there is no substantial issue here. It is simply a matter of how the usage of the *word* 'pain' should be refined by people who don't think that phenomenal and causal role concepts co-refer. Nor need this semantic issue exercise the rest of us, who think that these concepts do co-refer, and so can carry on using the word unreflectively in everyday contexts, as picking out that common referent.[4]

[4] According to my analysis, then, everyday usage must be committed implicitly either to materialism or to interactionism, given that it rests on an assumption of

My focus in this chapter will now be on the *phenomenal* concept of pain. My idea is thus that we should peel off a purely phenomenal element from the notions expressed by the everyday term 'pain'. This will be what we are left with, so to speak, when we have subtracted all psychological ideas of pain, all ideas of pains as things with certain characteristic causes and effects. Our task is to understand how this purely phenomenal concept is structured, and in virtue of what it has its referential power.

At some point some readers may be becoming uneasy. Could such a *purely* phenomenal concept really succeed in referring at all? Is not this idea dangerously close to a private language for mental 'givens'? I shall address this kind of concern at the end of this chapter. In particular, I shall there allow that there are some peculiar features of phenomenal concepts. Some self-ascriptive judgements made with phenomenal concepts have a special authority. Moreover, it is not automatic that someone who possesses a phenomenal concept will be able to communicate its content publicly. On the other hand, I shall argue that there is nothing in these features, when properly understood, to make us suspicious of phenomenal concepts. But all this is for later. Having flagged these issues, I shall now ignore them until the end of the chapter.

4.3 *Phenomenal Properties Provide their own 'Modes of Presentation'*

One sometimes sees it said that when phenomenal concepts refer to phenomenal properties, the latter provide their own 'modes of presentation' (e.g. Loar 1999). This thought is often associated with the claim, defended in the last chapter, that phenomenal concepts

phenomenal-psychological co-reference. This raises the question of which it is in fact committed to. But there is no reason to suppose that everyday thinking commits itself on this matter. It is enough that it assumes that phenomenal concepts refer to causally potent properties identified by phenomenal concepts. Given this much, everyday thinking has no particular need to exercise itself on the further question of whether those causally potent properties are regular material properties or made of some distinct mind-stuff.

refer to phenomenal properties directly, and not by invoking any further contingent properties of those referents. While I of course agree with this latter claim, I think that the frequently accompanying talk of 'their own modes of presentation' needs to be treated with extreme care.

At one level, the idea that phenomenal properties can provide their 'own modes of presentation' may simply mean that they do not have to be picked out via some *other* contingently connected property they possess. There is only one property in play when a phenomenal concept refers to a phenomenal property: namely, the phenomenal property itself. No further property mediates between referring concept and referent.

So far so good. This is just what I argued in the last chapter. But sometimes something more seems to be meant by 'providing their own modes of presentation', and here I think we need to be careful. I take it that a 'mode of presentation' is something like a Fregean sense, something grasped by the mind and with some kind of semantic power to latch on to a referent. The paradigm, perhaps, is where the mind is *already* able to think of some property, or combination of properties, Ø, and then uses this ability to construct a term to refer to the entity which possesses those properties ('the thing which has property Ø').

Now this Fregean picture of 'modes of presentation', I take it, indicates that we ought *not* to talk about phenomenal properties providing their 'own modes of presentation'. The idea we are working with is that phenomenal concepts refer to phenomenal properties directly, without mediation of any further properties. It would seem badly to misrepresent this idea to say that phenomenal properties provide their own Fregean modes of presentation. This suggests a picture whereby the mind somehow already has the power to think about some phenomenal property, Ø, and then uses this ability to form a mode of presenting that property ('the property which is property Ø', perhaps). But this makes little sense. If we already have the ability to think about the phenomenal property Ø, we don't need to construct some further mode of presentation to enable us to think about it.

However, there is a further circumstance in the offing which is capable of obscuring this point, and indeed of sowing great confusion about consciousness generally. Note that when we deploy phenomenal concepts, we also characteristically *instantiate* some version of the conscious property we are referring to.

This is most obvious with *introspective* uses of phenomenal concepts. When I pick out some aspect of my current experience introspectively ('this feeling . . . '), I *have* that feeling at the same time as referring to it. And a similar point applies to imaginative uses of phenomenal concepts. When I later think imaginatively about some earlier experience, like seeing red ('that experience . . . '), I won't actually have an experience of seeing red, but my experience is likely to bear some phenomenal similarity to the experience of seeing red—to be 'a faint copy', as Hume put it.

So in both cases the use of phenomenal concepts to refer to some experience will standardly involve the thinker actually having the experience itself, or a faint copy of it. Perhaps—though this is yet to be determined—we should think of this instantiation of the experience as literally *part* of the term the thinker uses to refer to that experience. And even if we don't go that far, we should certainly recognize that uses of phenomenal concepts will standardly be *accompanied* by versions of the experiences referred to.

Now, I take this feature of phenomenal concepts to be hugely important. Indeed, in Chapter 6 it will provide the crucial ingredient for my explanation of 'the intuition of distinctness'—that is, the widespread and well-nigh inescapable feeling that conscious and material properties must really be distinct. To give a very quick preview, in Chapter 6 I shall argue that it is easy to get confused by the fact that uses of phenomenal concepts involve the very phenomenal properties they refer to. For, when we compare phenomenal concepts in this respect with *material* concepts of conscious experience, which do not so involve the phenomenal properties they refer to, we note that there is a sense in which the material concepts 'leave out' the phenomenal properties. And from this it is very easy to slide, fallaciously, into the conclusion that material concepts cannot *refer* to phenomenal properties.

But all that is for later. Our current concern is not with confusions that might be generated by the special structure of phenomenal concepts, but with the analysis of how these concepts work in the first place. And here I think that the fact that uses of phenomenal concepts involve versions of their conscious referents is of no immediate importance. In particular, I do not think that this fact generates any immediate explanation of how those concepts refer to those experiences.

It is possible, however, to construe the idea that 'phenomenal properties are their own modes of presentation' as offering just such an explanation. The thought here would be that, in deploying phenomenal concepts, the mind is somehow in possession of an instance, or version, of the property being referred to, and that this in itself immediately accounts for the fact that those concepts refer to those properties.

I think this thought must be resisted. It betrays loose thinking about reference to suppose that concepts automatically refer to any properties that are involved in their deployment. Maybe the involvement of conscious properties in phenomenal concepts will turn out to be of some significance to those concepts' referential powers. But this can't be the whole story. After all, entities don't normally refer to themselves. So why should the presence of a conscious property in the mind automatically constitute a term which refers to that property? Still less do entities normally refer to whatever they might be 'faint copies' of. So again, merely instantiating a faint copy in imagination will not automatically constitute a term which refers to the original of that faint copy.

4.4 *World-Directed Perceptual Re-creation and Classification*

At this stage it will be helpful to turn away from phenomenal concepts, and consider some closely related mental powers.

Let me start by expanding my treatment of *imagination*. In my discussion so far, this has figured in what I have been calling 'imaginative uses of *phenomenal* concepts'. But a moment's reflection will show that this kind of contribution to thoughts *about* experiences is not the only way—or indeed the most basic way—in which imagination can contribute to thought. For we can also use imagination to think about non-mental things, like trees or houses or other perceivable objects.

For example, when I visually imagine the beach next to the house where I grew up, in Isipingo, South Africa, I do not normally do so in order to think about the *visual experiences* I used to have, but in order to think about the reefs, sandbanks, waves, and rock pools I so enjoyed. Again, when I visually re-create entering the Department Office in King's College London this morning, the normal upshot is that I think about the contents of the room, like desks, computers, and the departmental administrator, not about my matinal *visual experiences*.

My point here is that perceptual imagination is in the first instance a medium for thinking about the external world of macroscopic physical objects, and only secondarily a means of thinking about experiences themselves. To drive the point home, we need only consider the possibility of thinkers who have powers of perceptual imagination, but are incapable of thinking about experiences as such, who have no notion of mental states. I take it to be relatively uncontentious that some of our evolutionary ancestors must have been like this, if not some existing higher animals; moreover, it seems likely that some autistic people are also like this. Yet I also take it to be uncontentious that such thinkers could still use perceptual imagination to think about the world, to think about waves or rock pools, say, even though they can't use imagination to think about experiences. So parallel to—indeed, prior to—the use of perceptual imagination to think about experiences, there is a more basic use of perceptual imagination, to think about ordinary non-mental things.[5]

Now, a quite analogous point can also be made about the other use of phenomenal concepts, in introspective classification, as when we focus on some aspect of our current experience, and think 'this feeling . . .', 'this colour experience . . .' , and so on. Parallel to this

[5] Michael Martin (forthcoming) argues that perceptual imagining is always a matter of imagining experiences, on the grounds that perceptual imagining is always imagining from a point of view. I am not persuaded by this argument.

kind of introspective classification of experience stands ordinary *perceptual classification*. When I am looking at a visual scene, I will visually classify certain aspects of that scene. For example, when looking out to sea, I will see the waves *as* waves, say, and the seagulls *as* seagulls. Or in the office I might identify the new i-Mac as such, perhaps as a result of noting that it has one of those curious colours.

Again, such world-directed perceptual classification seems prior to introspective phenomenal classification. While some of the same sensory powers may be involved, their basic use is surely to think about the external world, rather than about experiences themselves. Consider again thinkers who are incapable of thinking about mental states as such. They can still use their powers of perceptual classification to think about the world, to think about waves and rock pools, even though they can't use them to think about experiences.

It is perhaps worth making clear that I take the underlying classificatory power here—perceiving *as*—to be a matter of perception rather than of judgement. I can see something *as* red, or *as* a cube, or *as* an elephant, even when I judge that it is not (because I know my visual system is being fooled in some way). As I am understanding it, the underlying power of perceiving *as* involves nothing beyond some kind of attention, wherein incoming stimuli are compared with some stored pattern, and a match between them is registered. Exercises of this underlying power can be taken up to form concepts which enter into full-fledged judgements (*this kind of seagull* is not found in Britain), but the power of perceiving *as* is in itself perceptual rather than judgemental.

4.5 *Perceptual Concepts*

In two sections time I shall consider the relationship between perceptual thinking about the non-mental world, on the one hand, and phenomenal thinking about experiences, on the other. But first let me say some more about the former world-directed powers.

To help keep things clear, I shall henceforth use the terms 'perceptual re-creation' and 'perceptual classification' specifically to refer to world-directed thinking using perceptual imagination and classification respectively, and I shall also talk about these as two uses of '*perceptual* concepts'. When I want to talk about the corresponding uses of phenomenal concepts to refer to experiences, I shall continue to speak of 'imaginative' and 'introspective' uses of *phenomenal concepts*.[6]

Note, to start with, that there is a question of whether we should talk about separate *re-creative* and *classificatory* perceptual concepts, as opposed to counting these as two uses of single perceptual concepts. This parallels the corresponding question which came up in connection with phenomenal concepts at the end of Chapter 2.

Thus, consider my perceptual concept of '*this* kind of bird', where I don't know anything else about the kind of bird in question, but can classify it visually, and can re-create it in visual imagination. Now, the classificatory power involved here seems dissociable from the re-creative power, and vice versa. It is easy enough to think of cases where one can classify something perceptually when it is present, but cannot re-create it in perceptual imagination. And a priori nothing seems to rule out the possibility of someone being able to re-create something perceptually, even though they are no good at picking it out when it is present (though this would admittedly be somewhat stranger). Given this possibility of dissociation, should we not recognize two different kinds of perceptual concept, re-creative versus classificatory, rather than one kind of concept variously deployed?

However, as with the corresponding question about phenomenal concepts, I shall not spend time on this issue. Once more, it seems a matter of description rather than substance. So sometimes I shall talk about perceptual concepts *simpliciter*, and at other times I shall

[6] Do *all* phenomenal concepts correspond to some mode of *perceptual* experience? While I am focusing on cases which do so correspond, I do not in fact wish to suppose that all phenomenal concepts are like this. As well as thinking phenomenally about vision, hearing, and the other familiar senses, we can also think phenomenally about pains, emotions, non-sensory thoughts, and various further states which everyday thought would certainly place outside the category of perception. Still, we can shelve this issue for the time being. In Chapter 7 I shall address the question of what, if anything, is characteristic of all the states we can think about phenomenally. But until then it will be helpful to concentrate on the paradigm cases of phenomenal concepts which do relate to perception.

distinguish between re-creative and classificatory uses of these concepts.

Another way in which perceptual concepts are like phenomenal concepts is in their dependence on prior experience. Possessing a perceptual concept of some entity normally requires that you have previously perceived that entity. You will not be able to classify something visually as a certain kind of bird, say, or as a certain colour, unless you have seen it before; nor will you be able to think about it using perceptual re-creation. This mirrors the point that the possession of a phenomenal concept requires that you have previously undergone the experience that the concept refers to.

Of course, the normal qualification is needed here, to allow that we can think perceptually about *complex* objects—red circles, say—that we have never seen before (provided, that is, that the requisite simple concepts have been derived from previous perceptions of red things and circular things). And a further qualification is needed, in the case of perceptual concepts, which doesn't apply to phenomenal concepts. For you can aquire a perceptual concept of a kestrel, say, even though you haven't perceived any instances, provided you have seen something which produces the same perceptual reactions—such as a picture of a kestrel, say, or a video of a kestrel in flight.

No doubt the explanation of the dependence on prior experience is the same as for the corresponding point about phenomenal concepts. When we deploy a perceptual concept to think about some non-mental entity, we will be activating some neural pattern. However, an original perception will have been needed to fix that pattern as something that can be so activated. As before, the brain needs an original from which to form the mould for further activations.

4.6 *How Do Perceptual Concepts Refer?*

Let me now focus on the referential powers of perceptual concepts. What makes it the case that my perceptual concept of '*this* kind of bird', say, indeed refers to the kind of bird in question? Clarity on this issue will be helpful in connection with various later issues, and in particular when we turn, in the next section, to the corresponding question about the referential powers of phenomenal concepts.

A first thought might be that perceptual concepts refer in virtue of the fact that exercises of them *resemble* their referents. I assume that this suggestion does not need to be taken seriously. It is true, to stick to the same example, that the bird in question will 'look' as things appear when we exercise visual concepts of 'that kind of bird'. (And, similarly, it will 'sound' as things seem when we exercise aural concepts of it, and 'smell' as things seem when we exercise olfactory concepts of it, . . .) But this is just the definitional truism that how the bird 'looks' to us is a matter of how we normally respond to it visually. To explain why those responses are about the bird in the first place would seem to require some more basic resemblance, between the bird *itself*, in abstraction from how it appears visually, and exercises of our visual concept of it. I know of no good way to make sense of this idea.

A second thought might be that perceptual concepts refer via descriptions which invoke phenomenal properties. Thus thoughts involving the perceptual concept 'that kind of bird' might be construed as equivalent to 'the kind of bird which produces *these visual experiences*'. And in general, perceptual concepts could be analysed as equivalent to 'the Ω which produces sensory experiences Ø'. This would then make the referential powers of perceptual concepts derivative from those of phenomenal concepts. Of course, these latter referential powers, of phenomenal concepts themselves, have yet to be explained. But the idea would be that, however they are explained, the referential powers of perceptual concepts would piggyback on them, in virtue of descriptive definitions along the above lines.

I have a simple argument against this suggestion, and in favour of the thesis that the referential powers of perceptual concepts must be independent of those of phenomenal concepts. Consider once more those beings (our evolutionary ancestors, or autistic people) who possess perceptual concepts, but no corresponding phenomenal concepts. These beings will be capable of thinking in perceptual terms about birds, trees, shapes, and colours—indeed, about anything perceptible—but will have no concepts of sensory experiences, or of perceptions, or of minds generally. Clearly the perceptual concepts of these beings cannot get their semantic power from descriptions framed using phenomenal concepts. For these beings will be able to deploy the perceptual concept 'this kind of bird', say, even though they are incapable of thinking about the *visual experiences* characteristically produced by that bird. So their perceptual concept must derive its referential powers from something other than their associating it with a description which uses phenomenal concepts.

Perhaps this is a bit quick. Once modern humans *are* sophisticated enough to possess phenomenal concepts, *then* they will certainly be capable of forming descriptions of the form 'the Ω which produces sensory experiences Ø'. So perhaps they will use such descriptions to replace, or transform, some of the simpler perceptual concepts deployed by their less sophisticated ancestors and other beings. Maybe this is what happens with concepts of 'secondary qualities', among people who are reflective enough to find reason to distinguish such qualities from other features of the external world.

Still, I find it quite implausible to suppose that *all* the perceptual concepts of normal modern humans have been so transformed. Apart from anything else, the possibility of beings who possess perceptual concepts, but no phenomenal concepts, shows that perceptual concepts are *capable* of referential powers in their own right, independently of any association with descriptions involving phenomenal concepts. Given this, it is hard to see what *motivation* there could be for transforming all perceptual concepts into phenomenal concept-involving descriptions. Maybe such transformations are justified in special cases, such as by the kind of considerations that might motivate demarcations of 'secondary qualities'. But this is no reason not to use untransformed perceptual concepts in their own right when such considerations do not apply.

This now returns us to the problem of explaining the referential powers of untransformed perceptual concepts. If they do not refer via descriptions relating them to phenomenal concepts, how exactly do they refer? I think the way forward here is to appeal to naturalistic theories of representation, in the style of causal or teleosemantic theories.

The simplest version of such a theory would be a straightforward causal account, according to which a perceptual concept refers to that entity which normally causes classificatory uses of that concept. For example, a perceptual concept might refer to some kind of bird because it is specifically birds of that kind which cause classificatory deployments of that concept.

The difficulties facing this simple causal story are well known. Most centrally, classificatory deployments of perceptual concept are often caused by things which the concept doesn't refer to. You can be fooled into visually judging that some kind of bird is present by mechanical birds, pictures, or tricks of the light. Yet your concept doesn't refer to a kind which includes these deceptive stimuli. This refutes the simple causal story. The trouble, in effect, is that the simple causal account of representation leaves no room for *mis*representation.

Teleosemantic theories promise to deal better with misrepresentation. Theories of this kind ask about the *purpose* of the perceptual concept, in a biological sense, rather than about its causes (with biological purposes cashed out aetiologically, in terms of histories of natural selection). The referential value of the concept can then be equated with those items which it is the biological function of the concept to track. Since concepts can malfunction, like other biological traits, it no longer follows that misrepresentation is impossible. Sometimes a concept will be activated when it is not supposed to be. (See Millikan 1984, 1989, Papineau 1984, 1993*a*.)

This is not the place to pursue details. In what follows, I shall simply assume that referential powers of perceptual concepts can be explained by some version of teleosemantics or, perhaps, by some revised version of the causal theory sophisticated enough to deal with misrepresentation (cf. Fodor 1990). Fortunately, none of the arguments which follow depend on the exact form of such a naturalistic version of representation. (Note, however, that such naturalistic theories portray perceptual concepts as referring *directly*, in the sense that the referential powers of these concepts do not derive from their association with further descriptions.)

4.7 *The Phenomenal Co-option of Perceptual Concepts*

I turn now to the referential powers of *phenomenal* concepts. Even if we assume that *perceptual* reference can be explained naturalistically, this does not yet tell us about *phenomenal* reference. We still need to explain the ability of phenomenal concepts to refer to phenomenal properties.

I take the following to be the obvious way of understanding phenomenal concepts. Originally there were only perceptual concepts. Our distant intellectual ancestors could classify things perceptually as birds, faces, colours, and so on. Moreover, they could use their powers of perceptual re-creation to think about such things even when they were absent. But they couldn't think about experiences.

Then they built on this basis to construct a practice for thinking about experiences themselves. A natural hypothesis is that they started to deploy concepts of the form 'the experience: ---', where the gap was filled by some actual perceptual classification or perceptual re-creation. By prefixing these perceptual states with the operator 'the experience: ---', they were able to generate terms which referred to the experiences themselves. Thus, for example, you might visually imagine something red, and, by prefixing this imaginative state with the experience operator, form a term apt to refer to a visual *experience* as of something red, as opposed to referring to a red surface. Or you might do the same while perceptually classifying some object as red, and again form a term with equivalent referential content, a term which refers to your *experience* of seeing something red, rather than to a red surface.[7]

It is plausible to regard the availability of these terms as part of the emergence of 'understanding of mind'. Human beings have a highly developed facility for thinking about their own and other

[7] As pointed out in n. 6, it should not be assumed that all phenomenal concepts correspond to *perceptual* states. In this connection, note that the model I have just sketched can obviously be adapted so as to allow phenomenal concepts of the form 'the experience: ---' in which the gap is filled by conscious states which are non-perceptual, such as emotions or conscious thoughts. Still, I shall continue to simplify the discussion in this chapter by sticking to the paradigm cases of perceptual experience.

individuals' mental states. The classic manifestation of this is their success on the 'false belief test', which requires the attribution of mistaken representations to other agents. Children are able to do this from the age of three or four onwards, though not before. It is unclear whether other animals can reason about minds to a similar extent.

Most discussions of 'understanding of mind', in this sense, have focused on the ability of humans to attribute *beliefs* and *desires* to each other, and to use these attributions to *predict behaviour*. In particular, there has been a detailed debate about whether we generate these predictions by *simulating* the decisions we would make if we ourselves had those beliefs and desires (the 'simulation-theory'), or whether we deduce the predictions from some general *theory* of the way beliefs and desires cause actions (the 'theory-theory'). (Cf. Davies and Stone 1995*a*, 1995*b*, Carruthers and Smith 1996.)

We need not enter into these issues here. For a start, my current interest in 'understanding of mind' is somewhat different from the standard one, in that I am concerned with the ability to think about conscious experiences, rather than about beliefs and desires. No doubt there will be some connections between these kinds of conceptual powers. In particular, it would be surprising to find thoughts about beliefs and desires in the absence of thoughts about experiences. Still, I have no special views about the way we refer to beliefs and desires, so can leave controversies on this matter to one side. My concern is solely to understand terms for phenomenal experiences.

Moreover, I have no need to take sides in the dispute between 'simulation-theory' and 'theory-theory'. In one sense, it is true, I have suggested that phenomenal thinking about experience involves a kind of simulation: I have hypothesized that mental terms for experiences are formed by adjoining an 'experience operator' ('the experience: ---') to an actual state of perceptual re-creation or perceptual classification. In this respect, I do think that phenomenal thoughts about experience involve a kind of simulation or instantiation of the experience being thought about. But this involves no commitment to the distinctive claims of 'simulation-theory'. It does not follow that any behavioural predictions drawn from such phenomenal thoughts must be generated by 'off-line' simulations of the way those experiences might lead to decisions. Even if you form terms for conscious experiences by activating some version of the experience itself, you may still reason with the terms so formed in an entirely theory-driven manner.[8]

4.8 *A Quotational Model*

The model I wish to pursue, then, proposes that phenomenal concepts are compound terms, formed by entering some state of perceptual classification or re-creation into the frame provided by a general experience operator 'the experience: ---'. For example, we might apply this experience operator to a state of visually classifying something as red, or a state of visually re-creating something red, and thereby form a term which refers to the phenomenal experience of seeing something red. Such terms will have a sort of self-referential structure. Very roughly speaking, we refer to a certain experience by producing an example of it.

It is worth emphasizing that I do not take the semantic power of these self-referential phenomenal terms to be self-explanatory. In section 4.6 I pointed to the possibility of a causal or teleosemantic account of the semantic power of perceptual concepts. I shall assume that the semantic power of phenomenal concepts is to be explained similarly: this power derives from facts about the causes or biological functions of the deployment of these terms.

[8] Why did phenomenal thinking first emerge in our evolutionary history? Presumably it served some useful function in the cognitive economy of our ancestors. Various conjectures are possible here: perhaps it first emerged in order to enable our ancestors better to predict the behaviour of others; or perhaps the point was to enable them to anticipate their own future experiences; or maybe the purpose was to facilitate reflection on the epistemological credentials of their own and others' beliefs. It is worth observing that on any one of these stories, or indeed on any similar story, our ancestors would have needed to associate their phenomenal concepts with psychological concepts which invoke the circumstantial causes and behavioural effects. For otherwise they would have been in no position to use their phenomenal concepts to help them predict the behaviour of others, or to anticipate which experiences would occur in which circumstances, or to figure out how experience will in certain circumstances mislead people as to the facts.

We should also note that phenomenal concepts are compound referring terms (composed of an 'experience operator' and a 'perceptual filling'). Any semantic theory will view the referential power of compound terms as deriving from systematic contributions made by their parts. Accordingly, a causal or teleosemantic account of phenomenal concepts will view the contribution of the parts to the semantic value of the whole as depending on the systematic contribution which those parts make to the causes or biological functions of the wholes they enter into. So there will be a story to be told about the contribution that the general experience operator 'the experience: - - - ' makes to the causes or biological functions of the phenomenal concepts it enters into, and a story to be told about the corresponding contribution made by the specific perceptual states that fill the space in this operator.

However, I shall not try to elaborate these stories here. Instead I shall simply proceed on the assumption that they are available. Some of the points which follow will depend on this assumption, and others will give hints about how it might be filled out. But, rather than get bogged down on this topic, which would raise both points of detail and general issues in the theory of reference, I have deemed it more fruitful to take these matters as read and press ahead.

It may be helpful to compare the model I am defending to the use of quotation marks. The referring term incorporates the things referred to, and thereby forms a compound which refers to that thing. Thus, ordinary quotation marks can be viewed as forming a frame, which, when filled by a word, yields a term for that word. Similarly, my phenomenal concepts involve a frame, which I have represented as 'the experience: - - - '; and, when this frame is filled by an experience, the whole then refers to that experience.[9]

From now on I shall assume this quotational model of phenomenal concepts. Let me now consider in rather more detail how it might work. It is not hard to see that the gloss I have offered so far needs some significant qualifications.

[9] I am slighly hesitant about highlighting this analogy, given that quotation raises its own puzzles (cf. Cappelen and Lepore 1997, Saka 1998). These are therefore likely to arise for phenomenal concepts too. Still, I don't think these puzzles invalidate any of the further claims I make about phenomenal concepts.

The most obvious is in connection with *imaginative* uses of phenomenal concepts. We are now assuming that these have the form 'the experience: ---', with the gap filled by an act of perceptual re-creation—visually imagining something red, say. And the suggestion I have made is that the resulting term will then refer to the experience which is 'quoted'—that is, to the experience which fills the gap in the experience operator. The trouble, however, is that the term in question does not standardly refer to this *imaginative* experience, but to the full-fledged experience of which it is a 'faint copy'. If I visually imagine a red square, say, and then think '*this* experience', I will not normally be thinking of the faint experience of imagining something red, but of the actual experience of seeing something red.[10]

To cope with this difficulty, the obvious solution is to have imaginative uses of phenomenal concepts referring, not to the imaginative experience that is 'quoted' itself, but to any experience that *resembles* it appropriately. To the extent that full-fledged experiences resemble the imaginative experiences that faintly copy them, this would then secure the desired reference to the full-fledged experiences.

This suggestion gains support from the model of linguistic quotation, and indeed from indexical constructions generally. When I use quotes to form a referring term, such as 'antidisestablishmentarianism', I will not normally be using this term to refer only to the word as written in lower case and in this particular typeface. Instead, I will be referring to a type which includes a wide range of possible inscriptions and sounds, with suitable linguistic or phonetic similarities to the exemplar within my quotation marks. Similarly with an ordinary indexical construction like 'that colour', used in connection with a particular sample. This will normally pick

[10] A complication. Even if this is the normal case, note that we *can* also use an imaginative act to refer phenomenally to the faint imaginative experience itself, rather than to the full-fledged experience it faintly copies. I *can* perceptually recall seeing something red, think 'the experience: ---', and thereby intend to refer to the faint recollective experience itself, rather than the actual experience of seeing something red. My hypothesis is that this involves some more or less explicit modification of the experience operator. My concept here is in effect 'the *imaginative* experience: ---'.

out a range of shades which resemble the sample, rather than the precise shade the sample displays. The same point applies, I would suggest, to imaginative uses of phenomenal concepts. The phenomenal concept will refer to a type of experience whose instances bear a certain resemblance to the 'quoted' exemplar.

It is true that I am here assuming an idea of *resemblance* among experiences. I am not, however, assuming that this notion need be explicit in the thinking of those who make an imaginative use of a phenomenal concept, any more than those who refer to the word type 'antidisestablishmentarianism' by quoting it need mentally articulate an idea of similarity between words, or than those who refer to *yellow* by indicating an instance need articulate an idea of chromatic similarity between colour samples. All that is needed is that subjects be disposed to use these terms to respond to such resembling instances in a uniform way, and perhaps that these dispositions have an appropriate history. On the causal or teleosemantic account of representation that I am assuming, it will be facts of this kind that determine the semantic power of terms which invoke appropriate resemblance to exemplars, whether or not the users of the terms articulate any ideas of such resemblances. In particular, it will be facts of this kind that will enable phenomenal concepts to refer to experiences which resemble 'quoted' exemplars appropriately.

A natural hypothesis about the responsive dispositions (the 'similarity spaces') which fix the semantic values of imaginative uses of phenomenal concepts is that they derive from pre-existing dispositions which similarly underly perceptual concepts. That is, subjects will treat experiences as phenomenally similar to the extent that they treat the features of the non-mental world that those experiences report on as perceptually similar. Their dispositions to respond to experiences uniformly will match their dispositions so to respond uniformly to the things those experiences are about. I treat a range of colour experiences, including experiences of perceptual re-creation, as similar precisely because I treat the features of the world that prompt those experiences as similar.

This discussion of 'similarity spaces' was motivated by the need to explain how *imaginative* uses of phenomenal concepts, which on the quotational account quote imaginative acts of perceptual re-creation, will nevertheless standardly refer to full-fledged perceptions rather than such recollective experiences themselves. Let me now consider how the quotational account fares with *introspective* uses of phenomenal concepts, and in particular whether its treatment of these uses will also need to appeal to resemblances among experiences, in the way we have just seen is necessary for imaginative uses.

Here things are not so clear-cut. On the quotational account, *introspective* uses of phenomenal concepts fill the frame 'the experience: ---' with perceptual classifications, rather than imaginative perceptual re-creations. While looking at something, I see it *as* red, say, or *as* a kestrel—and then I plug this state into the 'experience operator' to form a term which refers to the *experience* of seeing red or seeing a kestrel.

Now, states of perceptual classification do not seem to differ as sharply from the experiences thus referred to as do acts of perceptual re-creation. When I classify something *as* red, while looking at it, the experience this involves isn't a *faint copy* of the experience of seeing red. On the contrary, it is, if anything, a highlighting or intensification of that experience. The classification amplifies the underlying experience. In neural terms, we can usefully think of classification as occurring when some stored 'template' resonates with incoming signals, and thereby reinforces or augments them.

Given this, it seems that introspective uses of phenomenal concepts will actually *include* the experiences they refer to, in a way that imaginative uses do not. And, to this extent, there would seem to be no need in the introspective case to appeal to some *resemblance* between exemplar and referent to fill out the quotational-indexical story.

But perhaps the introspective case presents a converse difficulty. To the extent that the state of classification *intensifies* the underlying experience, it will itself be different from *un*intensified such experiences. It will be a *vivid* copy, so to speak, rather than a faint one. So, if we want it to be an exemplar for the full range, including the unintensified experiences,[11] there will again be a need to appeal to resemblance. On this suggestion, then, when we think 'the experience: ---', and fill in the gap with a state of perceptual classification, the resulting term should be understood as referring not just to experiences of the same vivid kind as the state of perceptual classification itself, but also to any non-vivid versions that resemble it appropriately.

Let me conclude this section by dealing briefly with a point that may be bothering some readers. At the end of section 4.3 I said that phenomenal reference does not arise simply because exercises of phenomenal concepts involve 'faint copies' of their referents, or those referents themselves. And at the beginning of this section I insisted that such reference is ultimately a causal or teleosemantic matter. But now I may seem to be taking these points back. For haven't I now argued that a phenomenal concept refers to whatever appropriately *resembles* the state that fills the operator 'the experience: ---'?

But this is no real conflict. I do indeed now want to say that phenomenal concepts refer to items that resemble their 'fillings'. But this doesn't yet tell us *why* phenomenal concepts so refer. And it would still be a mistake to view this as a direct upshot of the resemblance itself, supposing somehow that things automatically refer to whatever they resemble. Rather, the right answer is that it is an upshot of the causal or teleosemantic properties of phenomenal concepts. Phenomenal concepts refer to items that resemble their 'fillings' *because* applications of these concepts are typically caused by those items, or because it is the function of such concepts to track those items. We still need to appeal to a causal or teleosemantic theory of reference to explain why phenomenal concepts refer to what they resemble. Resemblance in itself does not explain anything.

[11] There is a question here about whether there is *any* perceptual consciousness in the absence of perceptual classification. Are our sensory states conscious at all, when they aren't being 'highlighted' by perceptual classification? Can we see consciously without seeing *as*? This issue will be discussed further in Chapter 7. For the moment, however, I shall simply assume that we can consciously see a kestrel, or the redness of something, even when we aren't visually classifying it *as* a kestrel, or *as* red.

4.9 *Indexicality and the Quotational Model*

At this stage I would like to return to the issue of how far phenomenal concepts are *indexical* constructions. I have presented a 'quotational model' of phenomenal concepts. But quotation can itself be seen as a special case of indexicality. Just as indexicals in general combine a descriptive specification with a directional indicator, so quotation marks specify a certain descriptive type (an inscription) and indicate a certain direction (inside the quotes). Given this, it will be helpful to say something more about the relation between phenomenal concepts and indexicality, especially in view of the frequency with which this issue is mentioned in the literature (cf. Horgan 1984, Bigelow and Pargetter 1990, Loar 1990, Rey 1991, Chalmers 1996: ch. 4, Tye 2000: ch. 2).

My view is that phenomenal concepts can indeed usefully be viewed as indexical terms, but that the indexical constructions involved are peculiar to the formation of phenomenal concepts, and cannot be assimilated to indexical constructions in use elsewhere.

It might not be immediately clear why phenomenal concepts are distinctive in this way. After all, on the quotational model, phenomenal thinking is still simply a matter of identifying an exemplar, and thereby referring to an item that resembles the exemplar. So why doubt that phenomenal thinking simply uses the same devices as any other indexical constructions that indicate some exemplar and thereby refer to a category that resembles it? Why should the phenomenal concept 'the experience: ---' work any differently from 'this bird', 'this car', or 'this colour'?

An initial reason for doubting this suggestion was provided by my initial discussion of phenomenal concepts and indexicality in Chapter 2. There I considered the possibility that Mary's acquisition of a new phenomenal concept was simply a matter of her now being able to ostend a relevant instance in her own experiential history ('this feeling'). But I showed that this suggestion would not do. We can't be ostending some actual past experience when we think *imaginatively* about experiences, for we may have forgotten how to locate any actual past experience of the requisite kind. Nor does it seem that we can effectively identify any particular feature of *present* experiences using an ordinary demonstrative like 'this feeling', for this will fail to specify which aspect of our current overall state of consciousness is being referred to.

Note how these difficulties are dealt with by the quotational model of phenomenal concepts. Imaginative uses of phenomenal concepts do not work by pointing directly to some past experience; rather, they 'quote' a *current* act of perceptual re-creation, and thereby refer to that experience which appropriately resembles that quoted exemplar. Again, introspective uses of phenomenal concepts 'quote' a state of perceptual classification, a perceiving *as*, and thereby refer to that experience which resembles that quoted exemplar.

So the simple indexical model considered in Chapter 2 can be dismissed. Still, this might not convince all readers that phenomenal concepts are *sui generis* in the indexical constructions they use. Maybe the exemplars involved in phenomenal conceptualization are required to be current states of perceptual re-creation or classification, rather than any old experiences. But, for all that, they still function as exemplars. So, once more, why deny that phenomenal thinking simply draws on the same devices as any other indexical reference by exemplification?

However, there is another feature of phenomenal concepts that is hard to square with this suggestion. Phenomenal concepts can only be formed using exemplars from the thinker's own mind. If phenomenal concepts were like other indexical terms which refer via exemplification ('this bird'), there would be no obvious reason why you should not indicate an experience in somebody else, and use this as an example with which to form a phenomenal concept. But I take it that any such construction would not yield a phenomenal concept. We can indeed form terms in this way. You tell your doctor about your uncomfortable foot, and the doctor responds '*That unpleasant feeling* is common in gout sufferers', intending thereby to refer to the category of experiences which resemble your own. But the doctor is not here using a phenomenal concept. For a phenomenal concept of gouty pain requires currently having the experience oneself, or being able to re-create it in perceptual imagination, whereas the doctor may never have experienced a gouty pain.

Phenomenal concepts are thus a peculiar species of indexical term. They can only be formed using exemplars from the thinker's own mind. This distinguishes them from other indexical constructions which use exemplification and resemblance. The analogy with quotation is close here. Quotation cannot be equated with the general possibility of referring to items as 'that inscription'. For this explanation leaves out the fact that the relevant inscription has to be placed *inside* the quotation marks. Similarly, phenomenal concepts cannot be equated with the general possibility of referring to things which resemble an exemplary experience, for this leaves out the fact that the relevant examples must be present *within* the mind of the thinker.

So there is perhaps something misleading about my representing phenomenal concepts as 'the experience: ---' plus some experiential filling. This could be taken to suggest that such concepts employ just the same devices as others that might be similarly expressed, such as 'that colour'. This would be wrong. Phenomenal constructions are peculiar, like quotation, and are not simply special uses of general indexical devices of exemplificatory reference. The experience operator I have schematized as 'the experience: ---' is not the same as constructions of the form 'that such-and-such'.

On the other hand, none of this is to deny that phenomenal concepts work in *similar* ways to other devices of exemplificatory reference. They may involve a *sui generis* construction, with strong restrictions on possible exemplars. But even so, they end up referring to things that resemble those exemplars, just as in other cases of indexical reference by exemplification. The analogy with quotation is instructive again. As I said above, quotation cannot be equated with 'that inscription', for this omits the requirement that the relevant inscription must be *inside* the quotation marks. But we can still understand quotation as a special kind of indexical construction: a quotational term indicates an exemplar *by* having it inside the quotation marks. Similarly, a phenomenal concept indicates an exemplar *by* operating on it with 'the experience: ---'.

To properly appreciate phenomenal concepts, we need to recognize that they draw on a *sui generis* construction, distinct from other indexical constructions. But we should also recognize that this is an indexical construction in its own right. Not only is this important for understanding the semantic workings of phenomenal concepts. It will also pay off when we come to consider various epistemological aspects of phenomenal thinking at the end of this chapter.

4.10 *The Causal Basis of Phenomenal Reference*

An objection to the kind of semantic story I have sketched might be raised as follows. 'Your account of the referential power of phenomenal concepts pays no attention to the distinctive *phenomenal* features possessed both by such concepts and by the phenomenal properties they refer to. The striking thing about these concepts and their referents is that they have a subjective nature. Exercising a phenomenal concept and experiencing its phenomenal referent are both like something—indeed, they are phenomenally like each other. Yet your account of their referential relationship makes no mention of this, but instead represents the relationship as an entirely causal or biofunctional matter. On your view, phenomenal concepts refer to phenomenal properties because of the causal or biofunctional connections between concept and property, not because of their shared subjective nature. But surely this is wrong. Surely phenomenal reference hinges on the *felt* nature of phenomenal referrers and referents, not on contingencies of their causal relationships or evolutionary history.'

We might make this objection graphic by considering a 'silicon zombie', who shares all your structural and historical properties, down to a level of fine detail.[12] It behaves in the same way as you, and has developed in the same environments. But the physical composition of its basic parts is different, involving a silicon-based organic chemistry rather than a carbon-based one. Now assume, for the sake of the argument, that consciousness derives from ultimate physical make-up, rather than from any structural or historical

[12] I owe this form of the objection to Ned Block. See also Balog 1999 on the possibility of 'quasi-phenomenal concepts' in a non-conscious being.

properties, in such a way that your silicon doppelganger is indeed a zombie, lacking any phenomenal properties.

Still, since the silicon zombie shares all your structural and historical properties, it will also share all your causal and biofunctional properties: these depend on organizational and historical matters, not on details of physical make-up. So, on my causal or teleosemantic approach to representation, it will follow that the silicon zombie will be your representational twin, even if not your phenomenal twin. So this zombie will be able to refer to its 'quasi-experiences' with its 'quasi-phenomenal concepts'. It will have concepts with the right causal or biofunctional qualifications to refer to the states which play experiential roles in it. But surely, the objection now goes, this zombie is not capable of the same kind of phenomenal reference as we are. If it lacks any subjective awareness, then surely it must lack the kind of mental grasp we have of our own conscious states.

It is important to focus on the right issue here. The question is not whether the silicon zombie lacks consciousness. This much we are currently supposing for the sake of the argument, though I shall query this supposition in a second. Rather, the issue is whether we can seriously suppose that the zombie's 'quasi-phenomenal concepts' will refer to its 'quasi-experiences', even though both these items lack the subjective phenomenality which constitutes conscious life in human beings.

My response is that, if we keep the relevant issue firmly in focus, there is no real difficulty presented by this thought-experiment. In general, there is every reason to suppose that referential relations are fixed by structural and historical matters, rather than by precise physical make-up. It would seem very odd to deny semantic powers to some alien creature, whose life is otherwise indistinguishable from ours, simply on the grounds that it has the wrong basic chemistry. Given this, I see no reason not to allow that the zombie doppelganger in particular would have the semantic power to refer 'quasi-phenomenally' to its 'quasi-experiences'. This power will be ensured if the relevant structural and historical requirements are met, and there is no reason to suppose that a variant chemistry would remove it.

Having said this, there is of course a rather different reason for doubting that the semantic power of 'quasi-phenomenal reference' can be found in the absence of genuine phenomenal subjectivity. For it is possible to doubt that the silicon 'zombie' would be a zombie to start with, on the grounds that the presence of appropriate representational properties may itself guarantee the presence of phenomenal subjectivity. This will follow if we adopt a 'representational theory of consciousness',[13] according to which conscious properties are constituted by representational properties. On any such theory, a representational duplicate will necessarily be a conscious duplicate, which would mean that there is no possibility of a silicon doppelganger who makes 'quasi-phenomenal references' and yet has no genuine subjectivity.

However, this line of thought is no objection to the account of phenomenal reference I am offering. For it does not dispute my thesis that the referential powers of phenomenal concepts derive from causal or biofunctional facts. Rather, it simply argues that, if a creature shares all our causal and biofunctional features, it must also share our conscious features. True, this line does imply that we can't have reference without subjective consciousness. But it doesn't argue this on the grounds that reference derives from consciousness, but rather that consciousness derives from reference.

4.11 *Phenomenal Concepts and Privacy*

I now want to address some worries about privacy. More specifically, I want to consider whether the close connection between phenomenal concepts and the first-personal perspective in some way casts doubt on the status of those concepts. I have in mind here some of the worries associated with Wittgenstein's 'private language argument'. There is no question of my dealing here with all the issues raised by Wittgenstein's argument. But I do at least hope to show that there is nothing unduly private about phenomenal concepts.

[13] Cf. Tye 1995, Dretske 1995. Representational theories of consciousness will be discussed in Chapter 7.

In the present section I shall consider a thought-experiment that provides some initial reason for thinking that phenomenal concepts are concepts in good standing, despite their constitutive connection with the first-person perspective. Then in the next two sections I shall consider whether first-personal phenomenal judgements are ever incorrigible, and whether any third-person phenomenal judgements about other people are ever well grounded.

The thought-experiment I wish to consider is one of the variant 'Mary' stories. Recall the case where Mary comes out of her house and is shown a sheet of coloured paper, but doesn't know which colour it is, in her old material terms. I take it that, even so, she therewith acquires a phenomenal concept of the experience occasioned by the paper. She can think about this experience, then and later, as 'the experience: ---', filling the gap with a state of perceptual classification or re-creation. And she can use the concept so formed to think thoughts with determinate truth conditions, as when she hazards 'I'll have that experience again before the day is out', or wonders whether or not 'That experience is the one normally produced by ripe tomatoes'.

Yet Mary's concept looks like a paradigm of the kind of thing Wittgenstein's private language argument is designed to discredit. For a start, Mary's use of the concept will not conform to any public criteria. Since there are no a priori links between phenomenal concepts and psychological ones, Mary's mere possession of the phenomenal concept will give her no idea of the characteristic external causes or behavioural effects of her new experience. Nor will she be able to communicate the thoughts that the concept enables her to form: if she coins a word ('qual', say) to express the concept, she will not be able to convey to her hearers what it means. Even so, I say, Mary's concept is a concept in good standing, in that it enables her to form thoughts with definite truth conditions. If this is so, then neither conformity to public criteria nor communicability can be essential to determinate thought.

There are two issues here: conformity to public criteria and communicability. Let me deal with these in turn. Public criteria first. If you think that representational content is somehow constituted by normative *rules* governing the deployment of concepts, then you may be inclined to resist the suggestion that Mary has a good concept even in the absence of public criteria. How could Mary's concept possibly have a determinate content, you will ask, if Mary is not sensitive to any normative principles tying its use to public criteria? However, I take this line of thought to cast doubt on the premiss that concepts require such normative rules. Since Mary clearly can think good thoughts with her new concept, say I, it follows that normative rules are inessential to representational content, at least the kind of rules that Mary lacks.

There are some large issues here, but my own view is that content does not derive from normative rules, but rather from the kind of non-normative natural facts invoked by causal or teleosemantic theories of representation. In so far as there are norms in the area of judgement, these *follow* from the prior naturalistic constitution of content, and are not a precondition thereof (cf. Papineau 1999). So, on my view, it is no deficiency in Mary's concept that she is not sensitive to any normative principles tying its use to public criteria. It is enough that her concept has appropriate causal or teleosemantic credentials, since this in itself will ensure that her concept refers determinately, and that judgements made by using it have definite truth conditions. (Of course, if we assume that it is 'correct' to make true judgements, and 'incorrect' to make false ones, then Mary will be subject to the 'norm' that she should judge truly; but this 'norm' doesn't require that Mary be sensitive to public criteria, only that her judgements have truth conditions, which requirement I take to be satisfied, for the reasons given.)

What about the incommunicability of Mary's concept? ('Well, let's assume the child is a genius and itself invents a name for the sensation!—But then, of course, he couldn't make himself understood when he used the word': Wittgenstein 1953: §257.) Again, since I take Mary to have a concept in good standing, I do not take communicability to be essential to determinate referential content. The thoughts Mary forms with the concept 'the experience: ---' have quite definite truth conditions, even if she can't communicate them to anybody else.

It would be worrying, however, if phenomenal concepts were *necessarily* incommunicable, if no one else could ever understand words used to express phenomenal concepts, even outside the special circumstances of our Mary thought-experiment. Certainly much of the argument of this book presupposes that such communication is possible. However, there is no great difficulty here. Our Mary may not immediately be able to make herself understood to normal English speakers with her term 'qual'. But nothing stops other better-placed speakers from communicating their phenomenal concepts, or indeed our Mary herself doing so, given more propitious circumstances.

What exactly is required to understand someone else's expression of a phenomenal concept? A weak requirement would be that you understand that the speaker is expressing a phenomenal concept, and that you know which experiential property it refers to. A stronger requirement would be that you be able to identify this experiential property via the same phenomenal concept, and not just via some material concept.

To see that there is no principled barrier to understanding expressions of phenomenal concepts, in either the weak or the strong sense, consider our Mary example again. To get a case of someone who satisfies the weak requirement, but not the strong one, let us suppose that Mary has a companion, Jennifer, who similarly has never seen colours but knows all about colour vision in material terms. Jennifer isn't shown the piece of paper that Mary sees, but is told in material terms that it is red. Then Jennifer, who of course knows that people acquire phenomenal concepts of experiences once they have had those experiences, will be able to understand Jane's 'qual' as expressing just such a phenomenal concept, and indeed one which refers to the experience caused in her by seeing something red.

To get a case of someone who can understand Mary in the strong sense, we can simply allow Jennifer to see the paper too. Then she too will acquire the phenomenal concept of seeing something red, and will thus be able to think about the experience referred to by 'qual' in the way in which Mary now does, and not just materially.

I take something like these two kinds of understanding to be part of our everyday appreciation of each other's talk about experiences. I explained earlier how I take an everyday term like 'seeing something red' to express both a psychological concept and a phenomenal concept. Given the involvement of psychological concepts here, everyday thinkers will be in a position to appreciate that other people have conscious experiences with certain characteristic causes and effects (such as being caused by ripe tomatoes, pillar-boxes, appropriately prepared pieces of paper, etc.). They will thence be able to infer that other people will form phenomenal concepts from those experiences. This will then put them in a position to form a weak understanding of other people's talk as expressing those phenomenal concepts.

In addition, normal everyday thinkers will themselves have the relevant experiences, and so will themselves have phenomenal concepts which refer to those experiences. So they will be able to use their own phenomenal concepts to think about the referents of the phenomenal concepts of other people, which will then also allow a strong understanding of other people's talk as expressing phenomenal concepts.

Of course, there are empirical presuppositions involved here. In particular, a strong understanding of other people's phenomenal talk will rest on the presupposition that they have the same experience as you in relevantly similar circumstances. Otherwise you will have no reason to suppose that the phenomenal concept you are using for the experience you have when looking at ripe tomatoes is the same as the phenomenal concept other people use for the experience they have in those circumstances.

Still, while it is indeed an empirical matter that different people have the same experiences in relevantly similar circumstances, there seems nothing especially worrying about this assumption. In particular, there seems no reason to doubt that the general run of such presuppositions can be confirmed by the kind of empirical evidence that will be discussed in Chapter 7.

This might seem a bit quick. Aren't I simply sweeping the traditional inverted spectrum problem under the carpet? How have I ruled out the possibility that, when some people look at ripe tomatoes, they have the phenomenal experience induced in the rest of us by looking at lush grass, and similarly with other experiences? I didn't stop to worry about this earlier, when discussing the everyday usage of experiential terms in section 4.2, not least because it is not clear that such inverted spectra would undermine everyday usage. (After all, nothing would show up in everyday discourse if spectra were inverted between people, as long as the different experiences in different people all played the same causal roles.) However, now I am attending explicitly to the question of whether we can have a strong understanding of each other's phenomenal concepts, the inverted spectrum issue does become relevant. For I would be wrong in thinking that the phenomenal concept you express by the words 'seeing something red' were the same as mine, if your spectrum were inverted with respect to mine.

The important point to appreciate here is that I don't take inverted spectra to be ruled out a priori, only a posteriori. It is quite *conceivable* that some people should experience something different from the rest of us when they look at red tomatoes. But in fact I take this hypothesis to be dismissible on empirical grounds. If we are materialists, then we have plenty of reason to suppose that the same material processes occur in different people when they look at ripe tomatoes, or are otherwise similarly stimulated; and we can anticipate that more detailed empirical research into brain mechanisms will confirm this. (Moreover, similar reasoning is available to interactionist dualists, and even to epiphenomenalists; which means that they too can reasonably take themselves to have a strong understanding of other people's phenomenal concepts. It is one thing to hold that some extra mind-stuff is activated when people have conscious experiences; it is another to hold that this mind-stuff will manifest itself differently in different people even though they are otherwise in similar circumstances. I take it that any good defence of dualist views will answer to principles of a posteriori theory choice, including a *ceteris paribus* preference for uniform causal mechanisms over heterogeneous ones. Given this, views that take different people to have the same mind-stuff in similar circumstances will surely be better supported than those that do not.)

4.12 *First-Person Incorrigibility*

Let me now turn to another worry connected with the apparent privacy of phenomenal concepts. This is the worry that judgements made using phenomenal concepts leave inadequate space for the possibility of error. When I judge phenomenally that I am in pain, there seems no room for me to be wrong. However, if there is no room for error here, can this be a genuine judgement? ('[W]hatever is going to seem right to me is right. And that only means that here we can't talk about "right" ': Wittgenstein 1953: §258.)

There are a number of issues raised by this worry. A first point to note is that nothing in my analysis of phenomenal concepts implies that they can only be used to describe the thinker's own conscious states. Phenomenal concepts may *incorporate* the thinker's own conscious states, but it does not at all follow that they cannot be used to describe the conscious states of other people. And there will certainly be plenty of room for error when subjects use their phenomenal concepts so to describe other people.

For example, if I know you are at the zoo, I might hypothesize, or positively judge, that *you* are having 'that experience' (where I re-create seeing something as an elephant) or 'this experience' (where I am myself currently seeing something as an elephant). And, more generally, there is nothing to stop me from forming any number of conjectures or beliefs about other people by using my own phenomenal concepts. The fact that these concepts are built from elements of my own experience does not mean that I cannot use them to characterize the experiences of other people, nor, obviously, that I cannot be mistaken when I do so (maybe you aren't looking at the elephant house, and so are not having 'this (elephant) experience', when I think you are).

Still, this initial observation on its own does not necessarily answer the underlying worry. Let us distinguish *first-person* uses of phenomenal conepts, which characterize the thinker's own experiences, from *third-person* uses, which characterize other people's experiences. The worry now would be that first-person judgements made using phenomenal concepts exclude the possibility of error in a way that is inconsistent with their status as real judgements.

Moreover, we can also now formulate a converse worry: namely, that third-person applications of phenomenal concepts suffer by epistemological comparison. (Can we ever really *know* what someone else is feeling?)

I shall consider the two sides of this asymmetry in turn. In the rest of this section I shall consider whether my overall story is threatened by the special authority it accords to certain first-personal uses of phenomenal concepts. In the next I shall consider whether it is threatened by any lack of authority it implies for third-person uses.

On the first question, let me begin by specifying the relevant notion of 'first-person use' more carefully. Not every use of a phenomenal concept to characterize a thinker's own experiences will possess a special authority. If I phenomenally judge that tomorrow I will have 'the experience: - - -' (seeing an elephant), on the grounds that I expect to go to the zoo tomorrow, this will have no greater authority than my phenomenal judgement, say, that *you* will see an elephant tomorrow. More generally, there will be many cases where I use phenomenal concepts to ascribe past and future experiences to myself on the basis of just the same kind of evidence as I might use to ascribe them to others. From now on I shall understand 'first-person use' as excluding these cases, and as referring specifically to those judgements about one's own experience that do not rest on such ordinary external evidence.

I want to consider two kinds of case under this heading: (a) phenomenal judgements which use the *same* state of perceptual classification or re-creation both to identify an experience and to classify it; (b) phenomenal judgements which use a state of perceptual classification to identify an experience and a *different* state of perceptual re-creation to classify it, or vice versa.

4.12.1 *Phenomenal Judgements which Use the Same State of Perceptual Classification or Re-creation to Identify an Experience and to Classify it*

Suppose I hear something as middle C (or see something as red, or see something as an elephant, . . .). Then I can use this current state of perceptual classification both to form a subject term which names this *particular* experience and to form a predicate term for this *type* of experience.

Now suppose that I use these terms to characterize such a current perceptual state as being of the relevant type. I judge that: *this particular experience is an instance of this type of experience.* Here there does indeed seem to be no possibility of error. I can't go wrong when I judge in this way that: *this* experience is an instance of hearing middle C, or this experience is seeing something red, or seeing an elephant, . . .[14] And the reason is clear enough. The type concept is formed from the selfsame experience that is identified as the subject of the judgement. This particular experience cannot fail to satisfy the type concept, since the type concept names the type which consists of experiences like your current one. In effect, the same experience features as both the referent of the subject term and the exemplar which gives the type concept its content. This thus removes any possibility of the judgement going astray.

Now, is there something amiss with this self-certification which my theory implies for such introspective phenomenal judgements? I do not think that there is. It seems to me that once we understand the mechanisms which ensure this, the consequence is unworrying. Consider an analogy. Given the way that 'I' and 'am here' work, there is no room for someone to be in error when they judge that 'I am here'. It falls out of the semantic constitution of these indexical terms that any judgements of this form must be true. Yet we do not on this account think that there is anything amiss with the semantics of 'I' or 'am here'. I take the same point to hold for phenomenal judgements like '*this* is an instance of an experience of red', and for not dissimilar reasons. Once we understand the semantic workings of phenomenal concepts, we can see why this kind of judgement cannot possibly go astray.

Now consider a slightly different case, where we use a state of perceptual re-creation, rather than of perceptual classification, both

[14] Of course, if you *report* your introspective characterization in public words, then your words may state something false, as a result of your using the wrong words to express your phenomenal judgement. But such errors of linguistic incompetence can arise for every species of judgement. The important point remains that first-person introspective judgements are peculiar in not admitting non-linguistic sources of error.

to name some past experience, and to form a type concept which we use to characterize that same experience. I visually imagine seeing an elephant, and then use this act of imagination (i) to refer to some particular past experience, and (ii) to characterize that past experience with a type-phenomenal concept formed from the same act of perceptual re-creation. 'That particular past experience was an instance of seeing something as an elephant.'

Perhaps in these cases of perceptual re-creation there is room for one kind of error that doesn't arise in the perceptual classification case. I might fail to name any particular past experience when I form the relevant subject term from my act of perceptual re-creation ('*that* particular past seeing of an elephant')—maybe because I have seen more than one elephant and can't distinguish the occasions, or for some such reason. But if we put such cases of reference failure to one side, then there is no remaining room for error, for the same reasons as in the case involving perceptual classification. Your subject term can't help but name an instance of the relevant type, if it names anything at all. For this type is picked out as consisting of instances which appropriately resemble your act of perceptual re-creation, while the particular experience which features as the subject of the judgement is picked out as a specific instance of just the same kind of resemblance. Once more, the semantic workings of phenomenal concepts remove any possibility of error.

4.12.2 *Phenomenal Judgements which Use a State of Perceptual Classification to Identify an Experience, and a Different State of Perceptual Re-creation to Classify it, or vice versa*

Now consider a rather different kind of case. I use a current state of perceptual classification to form the subject term, and a state of perceptual re-creation to form the characterizing type concept. 'My current experience [identified via a state of perceptual classification] is like *that* [and here an imaginative phenomenal concept is exercised].'

These phenomenal judgements do not enjoy the same guarantee as those considered in subsection 4.12.1. The subject term is here formed from a current state of perceptual classification, but is then characterized by a type concept formed using a quite different perceptual state, a state of perceptual re-creation. Since different states are used to form the two terms, there is no inbuilt connection. A judgement of this form could take a current visual state of seeing something as red, and characterize it falsely as like an imaginative re-creation of seeing something as green.

Still, perhaps another kind of epistemological guarantee is possible here. Consider the hypothesis that perceptual re-creation and perceptual classification may both be underpinned by the same mechanism.[15] Some kind of stored neural 'templates' may both (i) be reactivated in perceptual re-creation and (ii) used to establish matches with currently incoming stimuli in perceptual classification. If this is right, then it may mean that judgements which classify current perceptual classifications by imaginative phenomenal concepts are immune to error after all.

Look at it like this. Could you be mistaken in judging that a current state of perceptual classification—constituted by incoming stimuli resonating with some stored template A—is of a certain type—the type (faintly) exemplified by activations of template B? Well, maybe your making such a phenomenal judgement simply consists in A and B being one template rather than two. On this suggestion, phenomenally judging that your current perceptual state is of some imaginative phenomenal type would simply be a matter of the same stored pattern of activation being used *both* in your identification of your current perceptual state *and* in your typing it by your imaginative phenomenal concept. Conversely, to judge that your current perceptual state is *not* of some imaginative phenomenal type would simply be for two different stored templates to be involved here.

If this suggestion is right, then it rules out any possibility of error when you judge 'my current experience [identified as a state of perceptual classification] is like *that* [and here an imaginative phenomenal concept is exercised]'. The template identity which constitutes the judgement will simultaneously ensure that the judgement is true: since the same pattern of activation is involved

[15] Cf. the analogous suggestion about imaginative and introspective phenomenal thinking in section 2.9.

twice, the current experience referred to will inevitably be an instance of the type picked out by the characterizing concept.

Note how a similar analysis will apply if you refer to some particular past experience with the help of an *imaginative* act of perceptual re-creation, and then characterize it by a type-phenomenal concept formed from a current state of *perceptual classification*. 'That past experience [identified by an imaginative phenomenal concept] is like *this* [and here an introspective phenomenal concept is exercised].' True, as above, this form of judgement will allow reference failure in subject position, occurring when the imaginative phenomenal concept fails to pinpoint any particular past experience. But if we put these failures to one side, as before, and continue to view judgements of this form as a matter of the same stored pattern being used both to identify and to characterize the experience, then here too there will be immunity to error. The identity which constitutes the judgement will also make it true, for the experience will be referred to as an instance of the pattern used to pick out the characterizing type.

Obviously, these last remarks, about judgements constituted by template identity, are both speculative and underdeveloped. But, rather than try to elaborate them further here, let me simply settle for this rough indication of another way in which phenomenal judgements might enjoy a species of incorrigibility, alongside the more straightforward cases discussed in the previous subsection.

Readers of Wittgensteinian inclinations may still be feeling uneasy about the whole idea of incorrigible first-person phenomenal judgements about conscious experiences. But I think that this is quite the wrong reaction. Far from regarding these imputations of incorrigibility as an embarrassing corollary of my overall account of phenomenal concepts, I view them as a positive virtue. It is a familiar enough thought that introspective judgements about conscious mental states possess a peculiar kind of authority. I don't want to suggest that the remarks in this section have done anything more than point a way towards some possible understandings of such first-person authority. But I do at least take these possibilities to count in favour of my overall story, rather than against it.

4.13 *Third-Person Uses of Phenomenal Concepts*

Let me conclude this chapter by considering third-person uses of phenomenal concepts (where 'third-person' includes self-applications made on the basis of ordinary evidence). The worry about such uses is the mirror image of that addressed in the last section. Where first-person uses of phenomenal concepts can be thought to possess too much authority, third-person uses might be thought to possess too little. How can we ever know what other people are really feeling, you might ask, if all we have to go on is the external evidence of their circumstances and their behaviour? My account of phenomenal concepts implies a striking contrast with first-person judgements. Where first-person judgements are immune to certain kinds of mistakes, third-person uses seem hostage to the impenetrability of other minds.

I do not take there to be any substantial difficulty here. Maybe the authority of third-person phenomenal judgements is markedly inferior to that of first-person phenomenal judgements. But this doesn't show that there is anything wrong with third-person judgements. It is simply an upshot of the special immunity to error enjoyed by certain first-person judgements. Third-person judgements may lack this special immunity to error. But in this they are in the same boat as just about every other respectable claim to knowledge.

On one natural model, third-person applications of phenomenal concepts are *inferential*, or non-observational. You observe someone else's circumstances or behaviour, and then use various theoretical assumptions about the connection between phenomenal states and such evidence to draw a conclusion about the other person's conscious state. An alternative is to view third-person applications of phenomenal concepts as themselves directly *observational*, and so non-inferential. When confronted with someone in pain, I just *observe* the pain, as opposed to inferring it from the other's behaviour.

I see no need to decide between these two models. (My own view is that there are cases of both kinds.) Either way, third-person applications of phenomenal concepts will lack the special authority of first-person applications. They will be fallible in principle, even if not in practice, for all the reasons that make for fallibility in observation and theory-based inference in general. Yet it would be wrong to infer from this that there is something amiss with third-person phenomenal judgements. Unless we are going to dismiss all claims to knowledge which derive from observation or theory-based inference, which would be absurd, there is no reason to belittle third-person phenomenal judgements in particular.

Once more, empirical presuppositions are involved here. To apply a phenomenal concept third-personally in response to circumstantial or behavioural signs is to presuppose that these signs are reliable indicators of the referent of the relevant phenomenal concept. But, as before, there is no reason to regard such empirical presuppositions as worrying. Indeed, they have in effect already been discussed, under the heading of whether phenomenal concepts refer to the same things as associated psychological concepts. If such assumptions of co-reference can be warranted empirically, as I have suggested they can be, then so will corresponding practices which apply phenomenal concepts third-personally in response to the typical causes and effects invoked by psychological concepts.

Perhaps it is worth making clear that I am not supposing that any individual thinker needs to confirm the relevant empirical presuppositions personally, in order to be entitled to a phenomenal judgement based on behavioural or circumstantial indicators. I take it to be enough that these indicators are in fact reliable guides to the relevant phenomenal conclusions, whether or not individual judgers have checked that they are so reliable. In Chapter 7 I shall consider the kind of research which is capable of confirming such empirical presuppositions. But there is no imperative for individuals who make third-personal phenomenal judgements so to confirm them. It will suffice if they are disposed to make phenomenal judgements on the basis of signs which are in fact reliable indicators of the phenomenal facts.

Chapter 5
THE EXPLANATORY GAP

5.1 *Introduction*

Joseph Levine (1983, 1993) has argued that any attempt to construct materialist reductions of phenomenal states will leave us with an 'explanatory gap'.

Suppose we have some theory which identifies pain, say, with some physical property, like the firing of nociceptive-specific neurons in the parietal cortex. And suppose that this theory has all the empirical support it could have. As far as we can tell, pains occur when and only when parietal nociceptive-specific neurons are active. Moreover, these two properties seem to play exactly the same role in the causal scheme of things, to have exactly the same causes and effects. So, as materialists, we identify pain with the firing of nociceptive-specific neurons in the parietal cortex.

Even so, Levine argues, we will still lack any explanation of *why* nociceptive-specific neurons yield pain. There will still be a puzzle as to why it feels like *that* to have active nociceptive-specific neurons, rather than feeling some different way, or feeling like nothing at all. To be told that pains are always present when nociceptive-specific neurons are active is not yet to be told *why* those feelings should accompany those physical states.

The same point applies to materialist theories which identify pains with physically realized higher properties, rather than with the physical realizations themselves. Suppose we were to accept, again on the basis of the fullest empirical evidence, that pain is identical with the higher property of having some physical property which mediates between bodily damage and the desire to avoid the source of the damage. An analogous explanatory gap would still seem to remain. Why should possession of even this higher property feel like *that*? Again, we seem to lack any explanation of why the higher state should feel that way, rather than some different way, or no way at all.

The point generalizes. Take any phenomenal property C, and consider any theory that identifies it with some material property M. However well-supported this theory, it still seems to leave us in the dark as to *why* M yields C. Why does it feel like that, rather than some other way, or no way at all, to have M?

Levine argues that this explanatory gap is peculiar to attempted materialist reductions of *phenomenal* states. Materialist reductions in other areas of science do not leave us with any similar explanatory puzzle. Once water has been identified with H_2O, or temperature with mean kinetic energy, we do not continue to ask *why* H_2O yields water, or *why* mean kinetic energy yields temperature. And, in general, successful materialist reductions seem to explain the existence of non-phenomenal everyday kinds in a way that removes puzzlement.

So there seems to be something about phenomenal consciousness that materialist reductions cannot explain. Whereas other everyday kinds can be explained in material terms, consciousness seems to resist any materialist domestication.

In this chapter I shall examine this putative explanatory gap. My conclusion will be that there is nothing in it to worry materialists. The facts to which Levine draws attention do not amount to any substantial argument against materialism.

In showing this, I shall accept that there is indeed a kind of explanation which is not delivered by materialist reductions of conscious properties. And I shall also accept, with some qualifications, that this marks a contrast with materialist reductions in other areas of science. But I shall show that this is just what materialists should expect, at least those inflationist materialists who recognize distinct phenomenal concepts.

The reason we cannot give any materialist 'explanation' of why the brain yields phenomenal properties is not that these properties are non-material, where those studied in other areas of science are material. Rather, it is that phenomenal concepts are not associated with descriptions of causal roles in the same way as pre-theoretical terms in other areas of science. This means that it is possible to understand identity claims in other areas of science as involving descriptions, and so open to explanation by materialist reductions, in a way that is not open in the mind-brain case.

I shall deal with these issues concerning explanation in the next two sections. After that I shall turn to two related lines of argument which are sometimes brought against mind-brain reductions. In sections 5.4 and 5.5 respectively I shall consider the complaints that the relative non-explanatoriness of mind-brain reductions means (a) that they fall foul of the requirement that materialist reductions must follow a priori from the physical facts, and (b) that they lack the epistemological authority of reductions in other areas of science. I shall argue that neither of these complaints is justified either.

5.2 *Mark Twain, Samuel Clemens, and Intuitions of Gaps*

The best way to explain the basic point at issue is to compare mind-brain identities with identities involving proper names. Since proper names are not associated with canonical descriptions, there is no question of understanding proper-name identities as open to explanation. Similarly, I say, with mind-brain identities.

Consider this now well-known parable.[1] There are two groups of historians, one of which studies the famous American writer Mark Twain, while the other studies his less well-known contemporary, Samuel Clemens. The two groups have heard of each other, but their

[1] I got this story from Ned Block, and used it in when I first wrote about these matters (Papineau 1993*a*, 1993*b*). He can't remember where he got it from. (Cf. Block and Stalnaker 2000, and also Block 1978, where the moral that identities need no explaining was originally drawn.)

paths have tended not to cross. Then one year they both hold symposia at the American Historical Association. Late one night in the bar of the Chicago Sheraton the penny drops, and they realize that they have both been studying the same person.

At this stage there are plenty of good explanatory challenges that the historians might answer. Why did this person go under two names? Moreover, why did it take us so long to realize that Mark Twain and Samuel Clemens are the same person? But there is one request for explanation that they won't be able to answer, because it makes no good sense: why *are* Mark Twain and Samuel Clemens the same person? Once we realize that there is indeed only one person here, we can't sensibly seek to explain why 'they' are one person.

Phenomenal concepts, like proper names, refer directly, and for this reason mind-brain identities similarly raise no explanatory question. Let us suppose, for the sake of the argument, that we find out that pain is the firing of nociceptive-specific neurons in the parietal cortex. Then there are various explanatory challenges that we might take up. Why do we have two different kinds of concept (phenomenal and material) for this one property? And why is it so hard for us to recognize that there is just one property here (why is there so persistent an intuition of distinctness)? But there is one explanatory question we won't be able answer, because it makes no good sense: why *are* pain and nociceptive-specific neuronal activity the same property? Once we realize that there is only property here, we can't sensibly seek to explain why 'they' are the same property.

The point is that genuine identities need no explaining. If 'two' entities are one, then the one doesn't 'accompany' or 'give rise to' the other—it *is* the other. And if this is so then there is nothing to explain. It is possible to explain why *one* thing 'accompanies' or 'gives rise to' *another* thing. But you can't explain why one thing is itself.

Now, having made this point, I should immediately concede that I don't expect it to extinguish the underlying intuitions which fuel concern about the 'explanatory gap'. Let me go slowly here. I think that the Mark Twain example does provide a good model for the materialist reduction of phenomenal properties. And I therefore think that materialism leaves us in no more of an explanatory quandary than does the identification of Mark Twain with Samuel Clemens. But at the same time I recognize that this will strike many readers as unconvincing. Isn't it obvious that mind-brain materialism leaves something unexplained, in a way that the identity of Mark Twain with Samuel Clemens doesn't?

In defence of this last thought, you may want to point out that there is an obvious disanalogy between the two cases. Before the penny drops, some observant historians may become puzzled about the close proximity of Mark Twain and Samuel Clemens, and start wondering why these two people always turn up in the same places. Indeed, they may think up various possible explanations for this: perhaps the two are collaborating in some scheme, perhaps Clemens is following Twain, or whatever. However, once these historians realize that there is just one person at issue, their explanatory ambitions will dissolve. Their acceptance of the identity will quite nullify any desire for further explanation.

However, in the mind-brain case it seems quite otherwise. Even those, like myself, who are persuaded that the mind is identical to the material brain, will surely admit that they sometimes hanker for some further understanding of *why* brain activities should yield conscious feelings.

I concede that I sometimes find myself so hankering. But I do not think that this is because mind-brain materialism is somehow explanatorily inferior to the identification of Mark Twain with Samuel Clemens. Rather, it is simply because mind-brain materialism is so hard to accept in the first place.

The real fly in the ointment is the 'intuition of distinctness' that I have mentioned in previous chapters, and which will be the focus of the next. This arises quite independently of any questions of what materialism might or might not explain. Rather, it comes from a separate source, and seduces us into thinking that phenomenal properties must be distinct from material ones. So the underlying intuition here isn't that, *after* we have accepted materialism, then we will be left with some worryingly unexplained business. Rather, the intuition blocks our accepting materialism in the first place.

Of course—and this makes it hard to keep things straight—once we *have* been seduced by this independent intuition of distinctness into rejecting materialism, then we will indeed be faced with all kinds of unanswerable explanatory puzzles. If the phenomenal properties are distinct from material ones, then how come they always accompany each other? And how do the phenomenal properties get in on the causal act? And so on.

I am sure that it is questions like these which make people feel that there is some unanswerable 'explanatory gap' between brain and mind. But note how these explanatory problems presuppose that materialism is false. Correspondingly, *if* only we could convince ourselves properly to embrace materialism, we would be able to dismiss them as based on mistaken presuppositions, More generally, I maintain that, if we properly embraced materialism, then mind-brain identities would seem no more explanatorily puzzling than the identity of Mark Twain with Samuel Clemens.[2]

In support of this diagnosis, it is worth pointing out that the language used to posit an 'explanatory gap' often betrays an unacknowledged commitment to dualism. The problem is often posed as that of explaining how brain processes can 'generate', or 'cause', or 'give rise to', or 'yield', or 'be correlated with', or 'be accompanied by' conscious feelings. These phrases may seem innocuous, but they all implicitly presuppose that conscious feelings are some extra feature of reality, distinct from any material properties. And once we slip into this dualist way of thinking, then it is unsurprising that we find ourselves with unanswerable explanatory puzzles.

Given the points made in this section, the discussion in the rest of this chapter will have a slight air of unreality. In what follows, I shall consider how far different kinds of reductive theses can be held to explain identities. Mind-brain reductions will be argued not to yield

[2] Recall my response to Kripke's modal argument. There too I admitted an asymmetry: once Jane found out that Cicero = Tully, she ceased to think it possible that they could come apart; yet even convinced materialists like myself feel intuitively that zombies are possible. However, my diagnosis there (as here) wasn't that zombies are somehow intrinsically *more* possible that Cicero ≠ Tully (cf. materialism is somehow *less* explanatory than Mark Twain = Samuel Clemens). Rather, there is an independent source of intuition that persuades us that phenomenal properties are distinct from material ones, and *thence* we infer that this putative pair of properties could well come apart (cf. *then* we face an awful explanatory gap). The two cases are quite analogous.

any such explanations, unlike some scientific reductions, but like the 'reduction' of Mark Twain to Samuel Clemens. Further, I shall also consider whether this reflects badly on mind–brain materialism, and shall conclude that it does not.

However, I do not think, for the reasons just given, that this has anything much to do with the vivid intuition that materialism leaves us with a 'gap'. This vivid feeling is a consequence of the independently motivated intuition of distinctness, not of any *explanatory* deficiencies in materialism itself. To help keep things clear, I shall therefore avoid the term '*explanatory* gap' in what follows. This term only sows confusion. While there is indeed a strong intuitive feeling of a mind–brain 'gap', this does not derive from the relative non-explanatoriness of mind–brain reductions. Conversely, while mind–brain reductions are indeed less explanatory than many other scientific reductions, *this* isn't why we feel they leave us with a distinctive 'gap'. (After all, we don't feel this gap with other non-explanatory reductions, like that of Mark Twain to Samuel Clemens.)

5.3 *Reduction, Roles, and Explanation*

I have said that mind–brain reductions are less explanatory than characteristic reductions in other areas of science. It is worth considering carefully why this disanalogy should arise.

It might seem puzzling that there should be room for any kind of disanalogy here. In both the mind–brain and normal scientific cases, we start with certain *pre-theoretical* everyday terms, like 'pain' or 'thirst', in the phenomenal case, and terms like 'water', 'temperature', or 'lightning', in other areas of science. Then we establish empirically that these pre-theoretical kinds are coextensive with certain *theoretical* kinds. Pain coincides with *nociceptive-specific neuronal activity*, say, or water with H_2O. On this basis we conclude that the kinds are identical. But in both sorts of case this conclusion depends on the a posteriori discovery that the two kinds involved are found to be instantiated together. Given this, we might except both kinds of reductions to present themselves as matters of brute, unexplained fact. In the scientific case, as much as the mind–brain case, there is no a priori *reason* why the scientific and the everyday kind should go hand in hand. That is simply how the world turns out.

However, there is a further circumstance which arguably does distinguish the two kinds of case. The pre-theoretical kinds involved in scientific reductions will often be associated with descriptions of a causal role. Thus we can think of *water* pretheoretically as a liquid which is odourless, colourless, tasteless to humans, and *temperature* as a quantity which is raised by inputs of heat and causes heat sensations in humans, and *lightning* as a phenomenon which is produced by thunderstorms and illuminates the sky.

Suppose now that we have a materialist reduction of some such pre-theoretical kind. We discover that some physical property *is*, or *realizes*, the referent of the relevant pre-theoretical kind term. Then this reduction will in a sense allow us to *explain* such things as why the relevant kind is *water*, say. For the reduction will presumably show us how it is that this kind is colourless, odourless, and tasteless. Once we know that water is H_2O, we will be in a position to explain why it appears to humans in these ways. And so, if we understand the question 'Why is this kind *water*?' as the question 'Why is this kind colourless, odourless, and tasteless?', then we will have a satisfactory answer. More generally, whenever a material reduction tells us that some physical property is identical with, or realizes, some everyday kind, there would seem to be room for an explanation of why the relevant kind has the properties which constitute any associated role. In this sense we can thus explain 'why this quantity is *temperature*', understanding this as the question of why it is raised by inputs of heat and causes heat sensations in humans, and we can explain 'why this discharge is *lightning*', in the sense of explaining why it is produced by thunderstorms and illuminates the sky.

(In fact there is a complication here. For even if it is true that pre-theoretical kind terms like 'water' or 'temperature' are associated with descriptions of causal roles, it should not be taken for granted that these roles involve *physical* inputs and outputs. After all, 'colourless' and 'heat sensations' do not themselves look like terms of physics. And to the extent that such non-physical terms are involved, knowing about the physical nature of some kind need not immediately explain why it satisfies some associated role. Knowing about the *physical* workings of H_2O may leave us a long way short of knowing why it appears colourless to humans. However, let me skip over this issue for the moment. It will reappear at various points in this chapter.)

Let us now take it, for the sake of the argument, that standard scientific reductions explain why pre-theoretical kinds satisfy associated causal roles. We don't get the same result from reductions of phenomenal kinds to material kinds. This is because phenomenal concepts have no special associations with causal roles. When we think pre-theoretically of pain, using a phenomenal concept, we think of it in terms of *what it is like*, and not as a state with certain characteristic causes and effects. Because of this, a material reduction of pain will not have the same explanatory upshot as the reduction of water to H_2O.

Thus, suppose we are given that nociceptive-specific neuronal activity is identical to, or realizes, some phenomenal property, such as pain. Even given this, there will be no resulting explanation of why this physical activity is *pain*, analogous to the above explanations of why this liquid is *water*, or why this quantity is *temperature*, and so on. For these explanations hinged on the association of the relevant pre-theoretical kinds with causal roles. In so far as there are no causal roles associated with phenomenal concepts, no physical story is going to explain in any analogous way why certain physical activities yield conscious states. If we are not thinking of pain as something with certain physical causes and effects, but as something that *feels* a certain way, then we find ourselves quite unable to offer any explanation of *why* brains yield pains.

For the reasons indicated earlier, I do not think that this admission need embarrass us materialists about consciousness. Maybe we can't give any physical explanations of why brains generate feelings, in the way that we can explain why a certain liquid is water, or a certain quantity temperature. But this disanalogy between the mind–brain case and other scientific reductions does nothing to discredit mind–brain identities themselves. The source of the disanalogy is simply that phenomenal concepts are not associated with causal roles. So we have no option but to understand identity claims involving them as 'brute' identity claims. The only way of reading a mind-brain identity claim is as saying that one thing—a phenomenal property—is identical with another—a material property. Without any associated description of a causal role, there is no way of reading such a claim as stating the further fact that something satisfies that role. So there is nothing further to explain here, as there is in the scientific case. A mind-brain identity simply says of something that it is itself. Recall my analogy with Mark Twain. I say that once you really accept that pain, say, really *is* some material M, then you will see that this requires no more explanation than does Mark Twain = Samuel Clemens. Identities need no explaining.

This claim that identities need no explaining may seem to be belied by the scientific examples discussed a moment ago. Didn't I just admit that physics can yield an explanation of why a given liquid is water, or a given quantity temperature, and so on? If we can explain *these* identities, then why shouldn't we be able to explain mind-brain identities?

But the scientific examples are not really explanations of identities. We aren't explaining why this liquid is *water*—that is, why it is *the liquid* which in this world plays the role of being colourless, odourless, and so on. This would be to explain why this liquid is itself, which would be misplaced. Rather, we are explaining why this liquid is colourless, odourless, and tasteless. We are explaining why it satisfies the descriptions with which it is pre-theoretically associated. This is a perfectly good thing to explain, and I allowed above that physics can explain such things. However this is not a matter of explaining an *identity*—of explaining why some entity is itself— but rather of explaining why some entity possesses certain further attributes.

5.4 *Does Materialism Require the Physical Truths to Imply all the Truths?*

In the last section I said that pre-theoretical kind terms like 'water', 'temperature', 'lightning', and so on will standardly be associated with descriptions of a causal role. But I have so far in this chapter avoided committing myself on the further question of whether these associated roles also serve to *fix the reference* of these terms—that is, whether these terms name whatever natural kind plays the relevant causal role in the actual world. As I said in Chapter 3, I am not myself particularly convinced of this further claim. And it made no great difference in the last section, since the important point there was only that it is possible to understand questions about why something is water, say, as questions about why it is odourless, colourless, and tasteless, and not whether these descriptions also serve to fix reference.[3]

However, it will be convenient in the rest of this chapter to go along with this further reference-fixing assumption, and assume henceforth that most non-phenomenal pre-theoretical terms like 'water', 'temperature', and 'lightning' are not only naturally associated with descriptions of causal roles, but also have their referents fixed by these descriptions.[4] This is because there are two further arguments against mind–brain reductions that rest on this reference-fixing assumption. These two arguments start off from the explanatory asymmetry between mind–brain and other scientific reductions outlined in the last section. But they combine this asymmetry with the reference-fixing assumption to infer two further objections to mind–brain reductions: first, that these reductions do not satisfy the requirement that materialist reductions must follow a priori from the physical facts; second, that they do not have the

[3] Cf. Ch. 3 n. 4, where I pointed out similarly that the reference-fixing assumption is not crucial to Kripkean explanations of appearances of contingency: you could understand 'It is possible that H_2O might not be water' as 'It is possible that H_2O might not be odourless, . . . ', even if the reference-fixing assumption were false.

[4] In fact, as we have seen before, this causal role conception of reference can be taken in two different ways. We can take the relevant kind term to name some first-order property, but to identify it as a property which plays some specified causal role in the actual world. In Chapter 3 we assumed that 'water'—along with 'temperature' and 'lightning'—works like this, naming that first-order stuff which is colourless and so on in the actual world. But there are also role concepts that name higher properties, such as the property-of-having-some-first-order-property-which-plays-a-specified-causal-role. Dispositional terms, like 'soluble' or 'toxic', are arguably the paradigm case here. The points which follow in this chapter will apply equally to both understandings of role concepts.

epistemological backing that accrues to reductions found in other areas of science.

I shall show that neither of these arguments works. However, I shall not query the reference-fixing assumption about standard pre-theoretical terms, unconvinced though I am by this. Rather I shall show that, even if we grant this reference-fixing assumption, there is nothing in either of these further arguments to worry mind–brain materialists. I shall deal with the two arguments in turn, in this section and the next.

The first argument hinges on a particular characterization of materialism, which I shall call the 'a priori characterization of materialism' henceforth. According to this characterization, materialism is equivalent to the view that all truths—including all truths about the mind—follow a priori from the physical facts.[5] (Cf. Chalmers 1996, Jackson 1993, 1998.)

Why should anybody adopt this characterization of materialism? Well, focus on the reference-fixing thesis that standard pre-theoretical terms have their references fixed by descriptions of causal roles. This will make it a purely conceptual matter that *water*, say, is whatever actual stuff is odourless, colourless, and so on. Suppose now that the physical facts tell us that some actual physical property fills this role. It will then follow, without further ado, purely in virtue of the relevant term's a priori association with its causal role, that this physical property is, or realizes, the referent of the relevant pre-theoretical kind term. Once we are shown that H_2O is colourless, and so on, it follows a priori that H_2O is water. Once we are shown that mean kinetic energy is raised by inputs of heat, and so on, then it follows a priori that mean kinetic energy is temperature.

This doesn't of course mean that our overall reduction becomes entirely a priori. It is an empirical matter, which certainly does not follow from the definitions of 'water' and 'temperature' alone, that H_2O is colourless, odourless and tasteless, or that mean kinetic

[5] More precisely, this is normally argued to be a commitment of 'physicalism', rather than 'materialism'. But in this context 'physicalism' is normally understood as equivalent to my ' materialism', as including not just strict property identity of other kinds with physical properties, but also the possibility that other kinds are higher kinds which are realized by physical properties.

energy is raised by inputs of heat and causes heat sensations. The idea is rather that, once we have established these physical facts, then nothing *more* is needed, beyond conceptual analysis, to reach the reductive claims. In this sense, the reductive claims follow a priori from the physical facts alone. (Note how I am here implicitly assuming that 'H_2O is *colourless*' and 'mean kinetic energy *causes heat sensations*' are physical facts. This is the doubtful assumption flagged in the previous section. But once more we can let it pass. There are worse flaws in the a priori characterization of materialism.)

We can now understand the rationale for the a priori characterization of materialism. If there is a material reduction of *water*, then a full physical description of the world, plus the conceptual knowledge that water is the stuff that plays a certain role, will enable us a priori to identify which material kind reduces water. And then we will be able to read off any further truths involving water from our full physical description of the world. Moreover, the same will apply to temperature, lightning, and other pre-theoretical kinds. If some material stuff does reduce these kinds, then our conceptual knowledge of the pre-theoretically associated roles, plus full physical knowledge of the world, will enable us to identify the reducing stuffs, and thence read off any further facts involving the kinds.

Suppose now that such thoughts lead you to accept the a priori characterization of materialism. Then you may feel inclined to reject mind–brain materialism on the grounds that phenomenal facts *cannot* be inferred a priori from a full physical description of the world. This follows from the explanatory asymmetry outlined in the last section. There the lack of any canonical roles associated with phenomenal concepts precluded our understanding 'phenomenal pain' in a way which would allow us to explain why 'brains yield phenomenal pains'. Similarly here, the lack of any associated role to fix the reference of 'phenomenal pain' stops us from inferring facts about phenomenal pains a priori from physical facts about brains. Suppose you know everything there is to know about brain activities, and about the typical physical causes and effects of those activities. This won't enable you to figure out a priori that certain brain states *feel* a certain way. You won't be able to read off from all the physical facts involving nociceptive-specific neurons that it will *hurt* to have them active, or from all the physical facts involving visual area V4 that given activities there will amount to phenomenally *seeing something red*.

I agree entirely that phenomenal facts cannot be so inferred a priori from the physical facts, and thus that they violate materialism construed as the thesis that all facts must be so a priori inferable. But I don't take this to be a good argument against mind-brain materialism. I trust that it is clear how inflationary materialists like myself will respond here. We will simply reject the a priori characterization of materialism. We will deny that our materialism requires all truths to follow a priori from the physical truths.

Materialism would require this only if *all* concepts picked out their referents via descriptions of causal roles mediating between phsyical inputs and physical outputs, or were themselves physical concepts. If this were true, then all concepts of properties would indeed either be overtly physical—that is, pick out their referents *as* physical properties—or specify causal roles which mediate between physical inputs and outputs. And then, since materialism does require that all first-order properties are physical, it would follow that the full physical story will a priori fix the complete inventory of satisfiers of both physical and non-physical concepts—that is, a full inventory of all truths, however formulated.

However, the claim that all non-physical concepts refer via association with causal roles is precisely what inflationary materialism denies. Inflationary materialists take phenomenal concepts to refer directly, in their own right, and not via any specification of such roles. So inflationary materialists will see no reason to accept, even given their materialism, that the satisfaction of such concepts can be inferred a priori from any physical story, however full. As materialists, they will take the phenomenal concept of pain, for example, to refer to some material state. But since they also hold that this phenomenal concept has no a priori connections with causal roles of any kind, they will simply deny that its physical instantiation can be inferred a priori from any physical story.

In discussing Kripke's modal argument in Chapter 3, I had occasion to criticize 'the transparency thesis', according to which the truth of identity claims involving two directly referring terms must always be a priori knowable. This same dubious assumption seems to me to lie behind the characterization of materialism as requiring all truths to follow a priori from the physical truths. For this characterization simply presupposes the transparency thesis, when it assumes that any prima-facie non-physical concept which refers to a physical property must do so indirectly, via a descriptive association with some causal role. Inflationary materialists will simply deny the underlying assumption here. They will say that there is no reason why a priori distinct concepts should not both refer directly to the same thing, and in particular why a phenomenal concept should not refer directly to a physical property. That is, they will simply deny the transparency thesis. As before, this thesis seems quite unwarranted, a hangover from archaic assumptions about unmediated mental acquaintance which have been amply discredited by recent thinking about reference.[6]

5.5 *An Epistemological Gap*

Let me now consider the epistemological worry associated with the non-explanatoriness of mind-brain reductions. This is the worry

[6] Ned Block and Robert Stalnaker (2000) point out that something else is odd about the view that the physical truths imply all the truths. Even if all concepts other than directly referring physical concepts were role concepts mediating between physical inputs and outputs, would all truths really follow? What about the possibility that some role is filled twice over, by a non-material property as well as a physical one, as on the overdeterminationist 'belt and braces' view discussed in Chapter 1? Then there would be some truths—namely, about the non-material realizers of the role—which couldn't be derived from the physical truths and conceptual analysis. You might want to reply that such non-material fillers would themselves be inconsistent with materialism. Still, it is not clear that this serves. We can agree that materialism implies that there are no non-material properties. But does materialism imply that physics implies this? There are some tricky issues here, but I would say not. Physics itself does not plausibly imply anything about the non-existence of non-material stuff, even if we grant (which I deny) that materialism requires all concepts other than directly referring physical concepts to be role concepts mediating between physical inputs and outputs. Recall a point made in Chapter 1. The reason why materialists reject the non-material fillers of the 'belt and braces' view is not that they can read off their non-existence from physics and the analysis of role concepts, but rather that the general principles of scientific methodology advise against such extra fillers.

that the non-explanatoriness may remove our reason for *believing* in mind–brain reductions in the first place. Perhaps, this objection can concede, there wouldn't be anything wrong with material reductions of phenomenal states, if only we had reason to believe them. But how can we be in a position to believe them, if they require us to believe brute unexplained identities which cannot be derived a priori from the physical facts?

The thought here would be that any *epistemological* access to a mind–brain reduction must proceed via such an a priori demonstration (Levine 1993). Some causal role must be associated a priori with the reduced kind, and then the physical facts will show us how some physical property realizes that role. This, so the argument goes, is how we find out that H_2O is water, or that mean kinetic energy is temperature. Yet we have no such epistemological route to mind–brain reductions. As inflationary materialists admit, we can't similarly find out that some physical property realizes the role associated a priori with some phenomenal concept, since there are no such roles associated a priori with phenomenal concepts. So how can we find out that phenomenal states are material states?

A first materialist response to this challenge would be that this kind of a priori role-filling discovery isn't the only possible epistemological basis for believing in reductions. We can also have direct evidence for a reductive conclusion. Consider the straightforward causal argument for mind–brain identities adduced in Chapter 1. This owed nothing to any a priori analyses of the reduced phenomenal kinds as involving a priori roles. Indeed, for that matter, consider personal identities like Mark Twain = Samuel Clemens. These can't be epistemologically based on the uncovering of a priori role-filling, since there are no a priori roles in play here. Yet our knowledge of them seems none the worse for that.

Still, the objector might persist, knowledge of a priori role-filling is epistemologically crucial for *serious scientific reductions*. Maybe familiar everyday identities involving spatio-temporal particulars, like Mark Twain = Samuel Clemens, can be happily accepted on brute correlational evidence alone. But when it comes to the identifications of natural kinds with unfamiliar theoretical kinds, surely we need something more substantial, something that will enable us to understand *why* the theoretical kind realizes the everyday kind. And this will require knowledge of a priori role-filling, of the kind unavailable in the mind-brain case. This is why we find claims of mind-brain identity so unconvincing, by comparison with established scientific reductions. Or so this objection goes.

At this stage a different line of response is open to materialists. They can query whether established scientific reductions are deduced a priori from the physical facts in the first place, in the way the objection supposes. (Cf. Block and Stalnaker 2000.) So far I have not disputed this contention. But a moment's thought will show that it is highly dubious. Take water = H_2O. This has been known since the middle of the nineteenth century. But there was no possibility of any physical explanation of *why* H_2O is colourless, or has other water-identifying properties, until well into the twentieth century, with the advent of quantum mechanics. So the recognition that water is H_2O could not possibly have depended on any physical explanation of how H_2O realizes some a priori water role. Instead it must have been based on more direct evidence. Likewise, I would suggest, with many other scientific reductions. The recognition that lightning is atmospheric electricity derived from experiments, like Franklin's, which simply showed that electrical discharges occur when lightning does. It did not wait on any detailed physical explanation of *how* electrical discharges produce the effects associated with lightning.

A complication mentioned in the last two sections is relevant here. I have already pointed out that, even if our pre-theoretical concepts of water, lightning, temperature, and so on are role concepts, it doesn't follow that the relevant roles will be specified entirely in terms of *physical* inputs and outputs. And, on reflection, it is clear that any roles associated with these concepts will not be specified so purely physically. 'Colourless', 'odourless', 'flashing across the sky', and 'causing heat sensations' are not concepts used in physics. They are phenomenal or perceptual concepts, not concepts which appear in the vocabulary of physical science.

The assumption that all the relevant roles must be specified purely physically is presumably a consequence of the view of reference mentioned in the last section. This view—the one that the anti-materialists tried to foist on materialists—had it that all *non*-physical concepts must pick out their referents via purely physically specified causal roles. Once more materialists will simply reject this view. I have already emphasized the possibility of referring to material states directly, using phenomenal concepts. Also relevant at this stage is the possibility of referring to material states using *perceptual* concepts (such as 'colourless' or 'flashing across the sky'). All phenomenal concepts, and most perceptual concepts, play no role in physics itself, yet will be regarded by materialists as perfectly good ways of referring to real properties, and so as perfectly good ways of specifying the inputs and outputs of causal roles.

This means that, even if some pre-theoretical kind *is* identified by a role concept, there is no reason to expect the relevant role to be specified as mediating between physical inputs and outputs. The normal case is as likely to involve phenomenally and perceptually specified inputs and outputs as much as any physically specified ones. And this further means that, even if some kind is picked out by such a role, we won't be able to identify a realizer a priori *solely on the basis of physical information*. If the inputs and outputs aren't physically specified, physical information alone cannot tell us a priori how they are realized. Rather, at some point in accepting the reduction, we will have to embrace some brute phenomenal-physical identities, or brute perceptual-physical identities,[7] purely on the basis of direct correlational evidence.

I do not necessarily want to maintain that the identification of the fillers of a priori associated roles plays *no* part in establishing orthodox scientific identities. One possible route to the knowledge that a given pre-theoretical kind is identical with some material kind may be the discovery that a given physical property plays some causal role associated with that pre-theoretical kind. Still, in so far as this causal role is specified in terms of perceptual and phenomenal causes and effects, rather than purely physical ones, physics alone will not be able to identify the relevant physical realizer a priori. At some point the epistemological buck will have to stop. The physics and the conceptual analysis will fail to meet, and we will have to use

[7] In distinguishing these cases I am making the assumption, defended in section 4.6 above, that perceptual concepts typically refer directly in their own right, and not because of a priori associations with phenomenally specified roles.

direct correlational evidence to equate physical states with phenomenal or perceptual ones.[8]

The overall suggestion at issue in this section is that mind–brain identities must be epistemologically underprivileged by comparison with scientific reductions, since they cannot be derived a priori from physics. However, we have now seen that reductions in other areas of science cannot normally be derived a priori from physics either. Maybe typical scientific reductions make some use of role thinking. But in the end they rest on the acceptance of brute identities, just like mind–brain reductions.

I take it that no one will want to use this point to cast epistemological aspersions on standard scientific reductions. If we know anything, we know that water $=H_2O$. So, by the same coin, there is no reason for epistemological despondency about mind–brain reductions. They may require us to embrace brute identities, but so does the rest of science.

5.6 *Conclusion*

In this chapter I have conceded that mind–brain reductions do not explain why feelings exist. The physical facts do not explain *why* certain brain states constitute certain feelings. This is because phenomenal concepts are not associated with roles, and so there is no question of showing how certain physical entities fill those roles.

Still, I have argued that this does nothing to discredit mind–brain identities. Other familiar identities are equally inexplicable. For example, brute personal identities, like Mark Twain with Samuel Clemens, admit no explanation. Maybe scientific reductions, like that of water to H_2O are different; to the extent that 'water' is

[8] We can now also see that physical information alone will not *explain* why 'this liquid is water', if the causal role associated with the pre-theoretical concept 'water' involves perceptual and phenomenal causes and effects, as well as physical ones. At best, the physical facts will show that some physical kind mediates between the *physical* correlates of those perceptual and phenomenal causes and effects; but that these physical correlates constitute the relevant perceptual and phenomenal items will remain brute and unexplainable, given that these perceptual and phenomenal items are not themselves identified via causal roles.

associated with a role mediating between physical inputs and outputs, it is possible to read 'water' claims in a way that allows them to be explained physically. But even this contrast, as we have just seen, gives us no epistemological reason to distrust other reductions, like mind-brain reductions, which do not so generate physical explanations.

Some readers may feel that something must have gone wrong here. For surely, if we stop to think about it, and get away from the technicalities, there *is* something distinctively problematic about mind-brain identities. To return to the intitial concern about the 'explanatory gap', won't it always remain mysterious how brains give rise to pains, and colour experience, and all the rest of the rich phenomenal life we so enjoy? How could squishy grey matter possibly do all that? Everybody surely feels this puzzle. Yet we feel no corresponding mystery about Mark Twain = Samuel Clemens.

As I said earlier, I agree that there is something distinctively perturbing about the mind-brain case. But this doesn't show that there is anything wrong with the analysis in this chapter. Rather, it shows that something *else* makes us puzzled about mind-brain identities. As far as explanatoriness goes, mind-brain identities are no worse off than many other respectable identities. If explanation were all that mattered, we wouldn't find the mind-brain relation any more mysterious than Mark Twain = Samuel Clemens.

However, there is something else that matters: namely, the independent intuition of distinctness that I have mentioned before. This doesn't apply to Mark Twain = Samuel Clemens, which is why we have no difficulty with this identity. But it does apply to the mind-brain relation, and this is why we feel that it is different.

Chapter 6

THE INTUITION OF DISTINCTNESS

6.1 *Introduction*

Let me now focus on the intuition of distinctness itself. In my view, this is what makes the mind-body problem seem so intractable. Even given all the arguments, intuition continues to object to mind-brain identity. How can *pain* (which hurts so) possibly be the same thing as insensate molecules rushing around in nerve fibres? Or, to repeat Colin McGinn's question, how can our vivid *technicolour* phenomenology (our experience of reds and purples and so on) possibly be the same as cellular activity in *grey* matter?

In this chapter, I shall try to explain this intuitive resistance to materialism about the mind. I think there is indeed something special about the mind-brain relation. It generates this overwhelming intuition of distinctness. Even convinced materialists are likely to feel the pull of this intuition. I know that in my own case it continues to press, despite any amount of immersion in the arguments of the previous chapters.

However, I don't think that this contrary intuition discredits materialism, because I think it is mistaken. At the same time, I think that it is centrally important for materialism to recognize and explain this anti-materialist intuition. Materialism will remain unconvincing until this intuition is laid to rest. An intuition on its own does not amount to an argument. But it is a striking feature of the mind-brain relation that it does generate this contrary intuition, and a full understanding of the subject ought to explain why this is so. Even if materialism isn't mistaken, its defenders owe an explanation of why it should *seem* so mistaken.

6.2 *Is an Explanation Already to Hand?*

Some readers may feel that ample materials for such an explanation are already to hand. Consider the three anti-materialist arguments discussed in previous chapters—that is, Jackson's knowledge argument, Kripke's modal argument, and the argument from the non-explanatoriness of mind-brain reductions. I have argued that they do not establish their anti-materialist conclusions. Even so, might they not still be responsible for the *impression* that materialism is false? Even if the anti-materialist arguments are unsound, they aren't *obviously* unsound. So why not simply explain any intuitions of mind-brain distinctness as upshots of the persuasiveness of these arguments?

I have already had occasion to resist this suggestion in connection with the Kripkean and the explanatory gap arguments. Nor do I think that Jackson's argument on its own yields a satisfactory explanation of the crucial intuition. In this section and the next I shall confirm that these standard anti-materialist arguments do not explain the intuition of mind-brain distinctness. The implication will thus be that this intuition must have some other source, separate from the anti-materialist arguments considered so far.

Of course, given that there is this intuition of distinctness, it cannot help but lend apparent weight to the anti-materialist arguments. For the intuition will support the *conclusions* of these arguments, even if it stems from a quite different source. It will make the anti-materialist arguments seem more convincing than they deserve to seem, simply because it portrays them as leading to the truth.

But this is different from saying that the arguments explain the intuition. Indeed, it says precisely the opposite, since the intuition of distinctness will serve to *bolster* the arguments only if it gains credence from some independent source. There must be some other origin for the compelling intuition that the phenomenal mind is extra to the brain, if this intuition *adds* to the appeal of the standard anti-materialist arguments.

Let us first check whether the intuition of distinctness really does have an independent source, apart from the plausibility of the earlier anti-materialist arguments. The obvious way to demonstrate this is to show that analogues of these arguments apply equally well to cases where we do *not* find any corresponding intuition of distinctness.

I have already argued this in connection with both Kripke's modal argument and the argument from non-explanatoriness. Let me take these in reverse order. The last chapter showed clearly why non-explanatoriness cannot explain the intuition of distinctness. Materialism is indeed in one respect non-explanatory. Since phenomenal concepts do not allude to causal roles, there is no question of explaining how certain physical states play those roles. But other familiar identities are quite comparable in this respect. Identities involving ordinary names are brute identities, free of any allusions to causal roles. Yet we feel no persistent intuition of distinctness in these cases. So the mind–brain intuition of distinctness must depend on something else.

Now recall Kripke's modal argument. This started from the apparent contingency of the relation between phenomenal mind and brain, and sought to infer from this that phenomenal mind and brain must therefore be genuinely distinct. Now, the unsoundness of this argument is not at issue—in Chapter 3 we saw how materialists can answer Kripke. The current question is rather whether the Kripkean line of thought can account for the compelling *illusion* of mind–brain distinctness.

The example of Jane in Chapter 3 showed that it cannot. Jane picked up the names 'Cicero' and 'Tully' without knowing anything about them. The Cicero–Tully relationship thus initially struck her as brutely contingent. She thought that *Cicero might not be (or indeed is not) Tully*, quite analogously to the way in which you might think a brain might fail to be accompanied by a conscious mind. Yet, when Jane did discover that Cicero is indeed Tully, she had no residual intuition that Cicero *can't* really be Tully, akin to the intuition of distinctness found in the mind–brain case. Yet the Kripkean considerations apply to Cicero–Tully as much as to mind–brain cases: there are no associated roles to explain Jane's initial impression of Cicero–Tully contingency, any more than there are in mind–brain cases. So the persistent intuition of distinctness that we find in mind–brain cases must derive from something more than these Kripkean considerations.

6.3 *Does Conceptual Dualism Explain the Intuition of Distinctness?*

I turn now to Jackson's argument. When discussing this in Chapter 2, I made no explicit mention of the intuition of mind–brain distinctness. Still, some readers may feel that it is precisely Jackson's points that hold the key to this intuition. For, even if Jackson's argument fails to demonstrate the existence of non-material properties, it does at least establish that we have two very different ways of *thinking about* conscious properties, as either phenomenal or material. Perhaps this extreme difference at the conceptual level is what we need to account for the intuition of distinctness.

The thought here would be that this persistent intuition arises simply because our two ways of thinking about material properties are *so very different*. Presenting a feeling in phenomenal terms, as a feeling, and presenting it in material terms, as a material state, are qualitatively quite different mental acts. Jackson's argument forces us to recognize this striking difference between these two modes of reference. Maybe this difference is the reason why we find it so hard to believe that they pick out the same property.

In this connection, it is arguable that the proper-name and scientific cases do not display the same kind of conceptual dualism. While these may involve two personal names ('Cicero', 'Tully'), or two kind terms ('water', 'H_2O'), the names in these pairs do not display any striking qualitative difference. Given this, so the argument would go, we have no difficulty accepting that 'Cicero' and 'Tully', or 'water' and 'H_2O', can name the same things. It is easy enough to grasp the idea that two terms from the same general stable can name the same entity. By contrast, it may not be so easy to stomach an identity involving two radically different terms, one phenomenal and one material.

This suggestion receives some support from certain views about identity claims. According to Ruth Millikan (2000), for example, embracing an identity claim framed using two mental terms is effectively to start using the two terms as notational variants of each other. Where before you had two files of information, so to speak, one attached to each term, now you have merged the files, so only have use for one term.

However, perhaps such 'merging' is not so easy when the orginal terms are realized in radically different ways. Maybe the architecture of the brain somehow prevents the merging of a file attached to an ordinary material concept with one attached to a special phenomenal concept. You may convince yourself intellectually of the relevant identity. But somehow this intellectual recognition fails to produce the kind of cognitive simplification that comes with ordinary identities.

Now, I think this general line of thought points in the right direction. Later in this chapter I shall argue that the persistent intuition of mind-brain distinctness is indeed due to peculiarities of the dualistic conceptual structure by which we refer to conscious properties. But I do not think that it is enough just to point to the existence of this dualism. It is not just that we have two strikingly *different* ways of thinking about conscious properties. This alone does not explain the intuition of distinctness. Rather, the intuition derives from a special further feature of our dualistic conceptual structure. Or so I shall argue shortly.

But first I would like to refer back to Chapter 4 in support of my claim that conceptual dualism in itself is not enough to account for the intuition of distinctness. Remember my discussion of *perceptual* concepts. These were concepts associated with perceptual classification and perceptual re-creation, and referred to features of the external world, like birds, elephants, colours of objects, and so on.

Now, such perceptual concepts are themselves radically different from other ways of referring to things in the external world. To think of kestrels visually is a quite different mental act from thinking about them theoretically, say, on the basis of reading about their habits in an unillustrated book. This latter way of thinking would be available to somebody who had been blind from birth, yet such a person could not have a visual concept of a kestrel. As I pointed in Chapter 4, you can only have perceptual concepts for those simple things that you have perceived previously. So no amount of book learnin' will tell you how to think of something visually, if you have never had any visual experiences before.

Notice how this radical difference between perceptual concepts and other concepts means that Jackson's argument could as well be run with perceptual concepts as phenomenal ones. 'Consider an ornithological Mary who knows everything theoretical about kestrels, but who has been blind from birth, and so has no visual concept of a kestrel. Then her sight is restored, she sees a kestrel, and she acquires a visual concept of a kestrel.[1] Now she knows something she didn't know before. "*That* bird eats mice." So visual kestrels must be ontologically distinct from theoretical ones.'

Now, of course, this ontological distinctness does not follow, any more than it did for phenomenal properties and material brain properties. But my present concern is not with the soundness of Jackson's argument, which I take to have been discredited earlier, but rather with whether the kind of conceptual dualism that *is* established by Jackson's argument can explain the illusion of mind-brain distinctness. I take ornithological Mary to show that it cannot.

For ornithological Mary fits Jackson's argument just as well as the

[1] Will this Mary really now be able to form visual concepts on the basis of visual stimulation, if she has been blind from birth? Probably not, since her visual cortex won't have developed properly. But this doesn't undermine the substantial point that no amount of theoretical conceptualization will instil visual conceptualization. Note that a similar doubt could have been raised about the original Mary—would she be able to perceive red, if her V4 had never previously been stimulated by colours? Again, this quibble wouldn't have made any difference to the arguments. (The point of making ornithological Mary blind from birth, rather than merely deprived of kestrel experiences, is simply to bypass a number of irrelevant complications, such as the possibility of her constructing a complex visual concept of a kestrel out of simpler visual concepts which she does have.)

original Mary, and so displays a quite analogous conceptual dualism (though now it is a perceptual-theoretical conceptual dualism, by comparison with the original Mary's phenomenal-material conceptual dualism). This means that, if this kind of marked conceptual dualism is to account for the persistent intuition of phenomenal-material distinctness that we find in the mind-brain case, we ought also to find a persistent intuition of perceptual-theoretical distinctness in the ornithological case. In particular, any difficulty about 'merging files' embodied in different cognitive media ought to apply across the visual-theoretical divide, as much as across the phenomenal-material divide: given the close connection between perceptual and phenomenal concepts, any recalcitrance to merging found with phenomenal concepts ought to be equally characteristic of perceptual concepts.

Yet I take it that there will no such persistent intuition of distinctness for ornithological Mary. Once she finds out that her theoretical kestrels are the same as her new visual kestrels—the birds like *that*—she won't at this stage feel any residual intuition of distinctness. She won't continue to feel the pull of some thought that in reality there are two distinct kinds: the theoretical kestrels she has always known about and some extra, visual doppelgangers that she has now acquired access to.[2] (She won't, after all, find herself slipping into puzzlement about why non-visual birds always 'give rise' to, or 'are accompanied by', the visual ones.)

Yet, as I have said, conceptual dualism applies equally well to both Marys, in each case establishing a qualitative difference between her newly acquired concept and her pre-existing one, and erecting quite analogous barriers to 'file merging'. This shows that the mystery peculiar to the mind-brain relation must derive from something more than such conceptual difference *per se*.

6.4 *Nagel's Footnote*

In footnote 11 of 'What is it Like to be a Bat?', Thomas Nagel (1974) considers Kripke's challenge to mind-brain identity, and suggests

[2] Of course, she will continue to feel that there are two distinct ways of *thinking about* kestrels. But that is different.

that it might be met by distinguishing between 'perceptual' and 'sympathetic' imagination. As he puts it:

> To imagine something perceptually, we put ourselves into a conscious state resembling the state we would be in if we perceived it. To imagine something sympathetically, *we put ourselves in a conscious state resembling the thing itself. (This method can only be used to imagine mental events and states—our own or another's.)* . . . Where the imagination of physical features is perceptual and the imagination of mental features is sympathetic, it appears to us that we can imagine any experience occurring without its associated brain state, and vice versa. The relation between them will appear contingent, even if it is necessary, because of the independence of the disparate types of imagination. (My italics)

Now, there are a number of different thoughts in this passage. Some of them correspond to ideas I have already dismissed as insufficient to explain the intuition of distinctness. But the part I have italicized points to a rather different explanation of this intuition.

Let me go slowly. In the first instance, Nagel is simply pointing to the possibility of an inflationary materialism which recognizes two ways of thinking about the phenomenal/material realm. True, I have explained inflationary materialism in terms of a contrast between 'phenomenal' and 'material' thinking, rather than between 'sympathetic' and 'perceptual' thinking. Still, if we equate Nagel's 'sympathetic' with my 'phenomenal', and now include perceptual concepts as a special case of material concepts (cf. Ch. 2 n. 3 above), we can see him as advocating a special case of inflationism: materialists should appeal to the fact that we can think about conscious properties in two different ways—one phenomenal and one not.

Nagel then points out that this in itself offers materialists the wherewithal to explain why it *seems* possible for mind and brain to come apart, consistently with their commitment to this not being possible.[3] The two distinct concepts of conscious feeling can flank a term for non-identity, and there we are. Still, as I have argued previously, this in itself doesn't explain why intuition so *continues* to resist mind–brain identities, even in the face of strong arguments.

[3] This particular anti-Kripkean suggestion of Nagel's has been developed at length by Christopher Hill (1997). See also Hill and McLaughlin 1998.

Exactly the same explanation of an illusion of contingency would be available whenever we have two terms for one thing ('Cicero' and 'Tully'), yet in these other cases we are perfectly ready to accept the identity once we are shown the evidence.

Again, Nagel presses the point that these two ways of thinking about conscious properties involve two strikingly independent mental powers—sympathetic thinking on the one hand and perceptual thinking on the other. This corresponds to the thought considered at the end of the last section, that radical conceptual dualism *per se* might explain the intuition of mind-body distinctness. But this thought too was found wanting, given that analogous kinds of conceptual disparity seem not to generate any corresponding intuition of distinctness.

However, Nagel also says something further. In the part I italicized above, he observes that when we

> imagine something sympathetically, we put ourselves in a conscious state resembling the thing itself. (This method can only be used to imagine mental events and states—our own or another's.)

Though Nagel does not develop it, this seems to me the crucial point. Uses of phenomenal concepts *resemble* the conscious properties being referred to. Moreover, this kind of resemblance between concept and object is peculiar to uses of phenomenal concepts. So far, we have not appealed to this in trying to understand our peculiar attitude to mind-brain identities. I think that it holds the key to the intuition of distinctness.

6.5 *The Antipathetic Fallacy*

Let us now focus on the special feature of phenomenal concepts to which Nagel draws our attention—namely, that their uses *resemble* the conscious properties being referred to.

Consider the two ways in which phenomenal concepts can be deployed. They can be used *imaginatively* or *introspectively*. Both these exercises of phenomenal concepts have the unusual feature that we *use* versions of the experiences being referred to in the act of referring to them. When we deploy a phenomenal concept imaginatively, we activate a 'faint copy' of the experience referred to. And when we deploy a phenomenal concept introspectively, we amplify the experience referred to into a 'vivid copy' of itself.

In both these cases the experience itself is in a sense being *used* in our thinking, and so is present in us. For this reason exercising a phenomenal concept will *feel* like having the experience itself. When you think imaginatively about a pain, or about seeing something red—or even more, when you think introspectively about these experiences while having them—versions of these experiences themselves will be present in you, and because of this the activity of thinking phenomenally *about* pain or seeing something red will strike you introspectively as *involving* the feeling of these experiences themselves.

Now compare the exercise of some material concept which might refer to just the same conscious state. No similar feelings there. To think of activation of nociceptive-specific neurons, or of some-physical-state-which-arises-from-damage-and-causes-avoidance-desires, doesn't in itself create any feeling like pain. Or again, thinking of grey matter doesn't in itself make you experience seeing colours.

So there is an intuitive sense in which exercises of material concepts 'leave out' the experience at issue. They 'leave out' the pain and the technicolour phenomenology, in the sense that they don't activate or involve these experiences. Now, it is all too easy to slide from this to the conclusion that, in exercising material concepts, we are not thinking *about* the experiences themselves. After all, don't the material modes of thought 'leave out' the experiences, in a way that our phenomenal concepts do not? And doesn't this show that the material concepts simply don't refer to the experiences denoted by our phenomenal concept of pain?

This line of thought is terribly natural, and I think it is what lies behind the inescapable conviction that the mind must be extra to the brain. (Consider again how the standard rhetorical ploy juxtaposes phenomenal and material concepts: 'How can technicolour phenomenology arise from soggy grey matter?' 'How can this *panoply of feeling* arise from *mere neuronal activity*?')

However, this line of thought is a fallacy—indeed a species of use–mention fallacy, which elsewhere I have dubbed the 'antipathetic fallacy' (Papineau 1993*a*, 1993*b*). There is a sense in which material concepts do 'leave out' the feelings. They do not *use* the experiences in question—they do not activate them, by contrast with phenomenal concepts, which do activate the experiences. But it simply does not follow that material concepts 'leave out' the feelings in the sense of failing to *mention* them. They can still refer to the feelings, even though they don't activate them.

After all, most concepts don't use or involve the things they refer to. When I think of being rich, say, or having measles, this doesn't in any sense make me rich or give me measles. In *using* the states they mention, phenomenal concepts are very much the exception. So we shouldn't conclude on this account that material concepts, which work in the normal way of most concepts, in not using the states they mention, fail to refer to those states.

This then offers a natural account of the intuitive feeling that conscious experiences must be distinct from any material states. This feeling arises because we have a special way of thinking about our conscious experiences—namely, by using phenomenal concepts. We can think about our conscious experiences using concepts to which they bear a phenomenal resemblance. And this then creates the fallacious impression that other, material ways of thinking about those experiences fail to refer to the felt experiences themselves.[4]

6.6 *Do Phenomenal Concepts Resemble their Objects?*

The diagnosis I have offered appeals to Nagel's observation that 'when we imagine something sympathetically, we put ourselves in a conscious state *resembling* the thing itself' (my italics). Uses of perceptual concepts resemble the conscious feelings they refer to, and this is why other concepts of conscious states can seem to 'leave out' the feeling themselves.

[4] Brian Loar also points briefly to this possible explanation of anti-physicalist intuitions in his 'Phenomenal States' (1990: 90).

However, there is room to doubt Nagel's original contention. Do exercises of sympathetic imagination really resemble the experiences imagined? When I imagine a pain, for example, there is indeed something conscious going on. But surely this conscious occurrence does not feel the same as a *real* pain. It doesn't hurt, or make me desire its cessation. So why say it resembles the real pain?

The point generalizes. Even it is like something sympathetically to imagine experiences, these acts of sympathetic imagination are surely quite different phenomenally from the experiences themselves. Nobody is likely to muddle up an imaginative act with an actual experience. Far from resembling each other, they seem quite different in kind.

This is a reasonable challenge, which raises a number of interesting issues. Even so, I do not think it discredits my diagnosis of the antipathetic fallacy. Let me make two connected points in support of this claim.

First, even if *imaginative uses* of phenomenal concepts do not resemble the experiences imagined, these are not the *only* uses of phenomenal concepts. There are also *introspective* uses of phenomenal concepts. I classify some current experience as a pain, or as seeing something red, or as seeing a kestrel, and am thereby able to think about that experience.

Now, I take it to be uncontentious that *these* uses of phenomenal concepts resemble the experiences they refer to. As we saw back in Chapter 4, these introspective uses actually include the experiences themselves, while simultaneously highlighting or intensifying them. They are vivid copies of the experiences, rather than faint ones. So there is no doubt that an introspective phenomenal reference to a pain, say, will resemble a pain. Given that this referential act includes the pain, it will feel like a pain. It will hurt, and make me want it to go away.

So perhaps I should have restricted my diagnosis of the antipathetic fallacy to introspective uses of phenomenal concepts, and ignored imaginative uses. That is, I could have said that the impression that material thinking always 'leaves out' the experience itself arises specifically when we compare such material thinking with introspective phenomenal thinking. Since there is no doubt that *introspective* phenomenal references feel like the experiences referred to—after all, they include them—there is no room here to query the initial idea that phenomenal thoughts about experiences resemble the experiences themselves.

To this extent, sympathetic imagination is not the best case for my diagnosis of the antipathetic fallacy. Nagel's emphasis on imagination thus points us in somewhat the wrong direction.[5] A far more obvious source of antipathetic confusion is the resemblance of conscious states to introspective phenomenal thoughts, rather than to imaginative phenomenal thoughts.

Still, having made this concession, I do think I can after all defend the idea that imaginative phenomenal thoughts, as well as introspective ones, resemble their conscious objects, and that this also plays a part in seducing people into the antipathetic fallacy.

This brings me to my second point—namely, that the issue which matters here is whether normal people *take* there to be a resemblance between imaginative uses of phenomenal concepts and conscious feelings, not whether these judgements are defensible. If normal people judge that such imaginative uses 'include' their referents, then this will push them towards the antipathetic fallacy, whether or not these judgements pass any further tests of philosophical respectability. As soon as anyone *feels* that imaginative thoughts about experiences 'contain' their conscious objects, then they will be vulnerable to the fallacious corollary that material thoughts, by contrast, 'leave out' the conscious states.

I contend that people do make such subjective judgements about imaginative uses of phenomenal concepts. That is, I take it to be a common, everyday thought that such imaginative uses resemble the experiences imagined, even if it is possible to raise philosophical queries about such resemblances. Perhaps imaginative thinking about vision provides the clearest examples. Imagining seeing a red square resembles actually seeing a red square. Imagining seeing isn't

[5] Not that there is any definite indication that Nagel was trying to identify my antipathetic fallacy in his footnote 11. He certainly does not develop the diagnosis. On the other hand, there is nothing else in his footnote that hinges on issues of *resemblance* between conscious states and thoughts about them. This makes me think that the antipathetic fallacy must at least have been at the back of his mind.

exactly like seeing, of course, but there is an obvious sense in which such imagining and seeing are phenomenally similar from the subject's point of view. Nor is the phenomenon restricted to the visual realm. Even if imaginings of pains don't really hurt, they can share some of the phenomenal unpleasantness of real pains. An imagined pain may not be unpleasant in just the same way as a real one, but it can still make you feel queasy, or make you twitch, or make the hairs on your neck stand on end. Again, imagining tasting chocolate feels akin to actually tasting chocolate. Even if it's not as nice, it can still make your mouth water.

So I contend that, in cases like these, it is very natural for people to think that imaginative uses of phenomenal concepts resemble the experiences imagined. And then, I say, these thoughts will help push people towards the antipathetic mistake that material thoughts 'leave out' experiences.

Chapter 7

PROSPECTS FOR THE SCIENTIFIC STUDY OF PHENOMENAL CONSCIOUSNESS

7.1 *Introduction*

So far in this book I have argued for two main theses. First, we should be *ontological monists*. We need to identify conscious properties with material properties, if we are to have a satisfactory account of how conscious causes affect the physical world. Second, we should be *conceptual dualists*. We need to recognize a special phenomenal way of thinking about conscious properties, if we are to dispel the confusions that so readily persuade us that conscious properties cannot possibly be material.

The resulting version of materialism implies that there is much about consciousness of which we are a priori ignorant. Conceptual analysis alone is impotent to uncover the material essence of conscious properties. This is because phenomenal concepts have no a priori links with any material concepts, with concepts which pick out their referents *as* material properties. So, for each phenomenal concept, it will be an a posteriori matter, to be settled by empirical investigation, which specific material property it refers to.

True, I have already argued, in defending ontological monism, that each phenomenal concept must refer to *some* material property (where this general conclusion is itself an a posteriori claim, resting on the a posteriori thesis that conscious occurrences have physical effects). Still, as I observed in Chapter 1, this general conclusion by itself does not give us any specific knowledge of the referents of different phenomenal concepts. While it implies that any given phenomenal concept must refer to *some* material property, it does not tell us which specific material property that might be.

In this chapter I want to consider how empirical investigation might answer such more specific questions. How should science proceed if it wants to identify the material referent of our phenomenal concept of *pain*, say?

As it happens, I think that science can provide far fewer answers to such questions than many people suppose. There has been a great boom in 'consciousness studies' in the past few years. After many decades when consciousness was universally regarded as beyond the limits of science, the empirical investigation of conscious phenomena is now widely accepted as a scientifically legitimate enterprise. In itself this shift is clearly to be applauded; scientific investigation into conscious phenomena has uncovered many important and interesting facts, and will undoubtedly continue to do so. At the same time, the current enthusiasm for consciousness research has blinded many researchers to the real methodological pitfalls facing the empirical study of consciousness. It is a mistake to suppose that research into phenomenal consciousness can proceed just like other kinds of scientific research. Phenomenal concepts are peculiar, and some of the questions they pose for empirical investigation are peculiar too.

7.2 *The Limitations of Consciousness Research*

At first pass, it might seem obvious enough, in principle at least, how to identify the material referents of phenomenal concepts. Can't we simply ask subjects to tell us when they are in pain, say, and then check what is going on inside their brains? Of course there will be practical barriers to knowing about processes inside skulls; and we will also need sufficiently varied examples of pain to be confident that the brain processes we observe are characteristic of pain itself, and not of some wider or narrower category.[1] Still, these look like standard scientific problems, which can in principle be overcome. So, in general, why can't we identify the material nature of any phenomenal property simply by investigating which material processes occur when that phenomenal property is instantiated in ordinary human beings?

However, it will turn out that this strategy is limited in essential ways. The trouble is that research involving ordinary human beings will fail to pinpoint the material referents of our phenomenal concepts, even given epistemologically ideal circumstances. However much we know about our cerebral innards, and however varied the examples of human pain we are given, there will still be a number of distinct material properties which this sort of research will be unable to decide between as the material essence of pain.

The problem is highlighted when we consider whether other animals, or future computer robots, or possible extraterrestrials, have experiences like phenomenal pain. The problem is that these creatures may share some of the material properties which are characteristic of pain in humans, but not others. So standard empirical research involving ordinary humans will fail to tell us whether these other beings can feel pain. Since such research cannot pinpoint any precise material property as the essence of pain, it cannot tell us exactly what is materially required for non-human creatures to feel pain.

It may seem as if this difficulty could in principle be overcome by appealing to a yet more extensive database of examples, including non-human creatures alongside ordinary human subjects. This promises to give us cases which display some of the material candidates for phenomenal pain, but not others. So it seems that we ought to be able to pinpoint the right material referent, by checking whether or not phenomenal pain is still present in these cases.

But I shall show that this strategy will not work. The problem here is principled, not practical. The barrier is not simply that of finding

[1] For a sensitive discussion of the issues involved here, see Frith *et al.* 1999.

(or engineering) examples which dissociate the material properties that invariably co-occur in ordinary humans. Rather, it is that such examples won't help us, because the methodology of consciousness research breaks down in the face of such cases. Even if we did have examples of the required kind, our methodology would have no grip on them, and would fail to deliver the answers we want.

Ned Block has argued that this problem is the Achilles' heel of inflationist materialism: if you introduce phenomenal concepts, you won't be able to identify their material referents, and so won't be able to decide whether or not non-human creatures satisfy these concepts (Block forthcoming). I agree with Block that this indecision is a consequence of the inflationist recognition of phenomenal concepts. However, I don't agree that this represents some kind of deficiency in inflationist materialism. In my view, it is indeed not always possible to answer such questions as whether octopuses, say, or advanced computer robots, or Proxima Centaurians, can feel phenomenal pain. So I regard it as a virtue of inflationist materialism that it implies such questions may be unanswerable.

Why exactly are such questions unanswerable? One possibility is that questions about phenomenal consciousness always have definite answers, but epistemological obstacles bar our access to them. This would indeed be puzzling, given materialism. If phenomenal properties are determinately material properties, then why shouldn't we be able to find out about their material natures? But there is another possibility. Perhaps the reason we can't always answer questions about phenomenal consciousness in non-human creatures is that our phenomenal concepts are *vague*.

I shall be arguing for this analysis. There are no definite facts of the matter about the applicability of phenomenal concepts in doubtful cases, and this is why we can't always give definite answers to questions about phenomenal consciousness in non-human creatures. My reason for this imputation of vagueness is not simply that we sometimes find ourselves unable to provide definite answers to the questions at issue. I am not guilty of the verificationist sin of inferring an indefiniteness of answers immediately from the undecidability of questions. Rather, I shall argue that there are independent reasons, relating to the special constitution of phenomenal concepts, why such concepts are vague in certain dimensions. The problem has nothing to do with our epistemological limitations. Not even an omniscient God could tell whether an octopus feels phenomenal pain, for the same reason that he couldn't tell whether I am bald.

At first sight, it may seem very odd to hold that questions about phenomenal consciousness are vague. Surely, we feel, there is a fact of the matter about whether a octopus feels like *this* (and here we 'quote' some instance of real or re-created pain). But the position is not so odd, once we become clear about what is being claimed. My thesis will not be that there is anything vague about how it is for the octopus itself. Rather, the vagueness lies in our concepts, and in particular whether such phenomenal concepts as pain draw a precise enough boundary to decide whether octopuses lie inside or outside. More generally, I shall argue that all our phenomenal concepts are too vague to draw sharp lines, once we extend them beyond their everyday range of application.

7.3 *Phenomenal and Psychological Research*

It is worth emphasizing that my concerns in this chapter are entirely to do with research into the material referents of *phenomenal* concepts, and not with other kinds of psychological research. As I explained in Chapter 4, I take phenomenal concepts to be only part of what is expressed by everyday mental terminology, like 'pain', 'seeing something red', 'hearing middle C', and so on. As well as expressing phenomenal concepts, such terms also express *psychological* concepts—that is, concepts associated with causal roles mediating between canonical perceptual inputs and behavioural displays.

To some extent, the difficulties I shall be concerned with in this chapter are obscured because research into the material referents of *phenomenal* concepts is not always distinguished clearly from the quite different, and methodologically unproblematic, enterprise of research into the referents of *psychological* concepts. With research of this latter kind, we can tell a familiar story: it is an a priori matter which causal roles are associated with which mental terms; everyday observation can then show us when these roles are satisfied, and in which creatures; and more detailed scientific investigation can then uncover the physical states which realize these roles in different creatures.

I do not say that these matters are trivial. No doubt there are serious challenges involved in delineating an interesting psychological concept of 'object recognition' or 'causal reasoning' and then figuring out which creatures satisfy these concepts, and which physiological mechanisms enable them to do so in each case.

Still, the difficulties involved in research into phenomenal concepts go beyond any of these psychological issues. Psychological work on 'object recognition' or 'causal reasoning' may be intellectually challenging, but these topics raise no deep philosophical difficulties. The problems posed by phenomenal concepts, by contrast, transcend these relatively mundane matters. This, I take it, is why we feel so pressed by the question of whether it is ever like *this* (and here we quote some experience) for octopuses. It would be quite mysterious why such questions should agitate us so if all they involved were issues about specifying causal roles and figuring out how such roles are filled in octopuses.[2]

This dissociation between phenomenal and psychological research is of course a corollary of the overall argument of this book. I have stressed throughout that phenomenal concepts must be distinguished from psychological and other material concepts, and that there are no a priori connections across this divide, Given this, it is no surprise that empirical research into psychological concepts should prove impotent to decide phenomenal questions.

[2] Or, indeed, it would be mysterious why such questions should agitate us so if all they involved were the issue of whether psychological role concepts refer to roles or realizers (cf. Ch. 2 n. 2, Ch. 4 n. 1 and Ch. 5 n. 4). There is of course a great deal of literature on whether pain and other mental states are functional roles or physical realizers. But this literature does not normally distinguish between psychological and phenomenal concepts. In my view, the interesting question here is whether phenomenal concepts refer to roles or realizers; indeed, this question will figure prominently below. Once this question is separated off, the remaining issue about psychological concepts seems unexciting.

Discoveries about psychological pain carry no immediate implications concerning the presence of phenomenal pain.

Recall the knowledge argument discussed in Chapter 2. As we saw, this failed to disprove ontological materialism, but it did establish conceptual dualism. In so doing, it provided a graphic demonstration of the impotence of psychological research to decide phenomenal questions. You can know as much as you like about canonical octopus responses to stimuli, and about physiological processes inside octopuses, and you still won't know whether the octopus feels like *this*. That is, no information about the material realizations of psychological concepts will tell us when phenomenal concepts are satisfied. If we want to find out about the referents of phenomenal concepts, we will need to do something different from simply figuring out how given causal roles are satisfied in different creatures.

I take it that much current consciousness research is designed to do precisely this. In any case, this is the kind of research that I shall be concerned with in the rest of this chapter. This is not of course because there is anything wrong with research into psychological concepts, but simply because it is phenomenal research that poses the more fundamental philosophical puzzles.[3]

7.4 *Subjects' First-Person Reports*

There is a distinguishing mark of research that is designed to identify the referents of phenomenal concepts. This is the crucial role that it accords to subjects' first-person reports on their phenomenal states.

Thus consider the standard strategy adopted in paradigms of recent research into consciousness. Experimental subjects are presented with certain stimuli, or asked to perform certain tasks. At

[3] David Chalmers (1996) distinguishes the 'hard problem', of figuring out what phenomenal concepts refer to, from the 'easy problem', of showing how various creatures satisfy various causal roles. The points made in this section endorse this distinction. Of course, I disagree with Chalmers's later claim that the right solution to the 'hard problem' is that phenomenal concepts refer to non-material properties; but I agree with him that the 'hard' problem is distinct from the 'easy' one.

the same time, researchers seek to figure out what is going on inside their skulls, using traditional techniques like electroencephalography (EEG), or more recent functional imaging techniques such as Positron Emission Topography (PET) and Magnetic Resonance Imaging (MRI),[4] or indeed simply by noting that subjects have suffered various kinds of brain damage. And then the experimenters will *ask* the subjects what they *experienced* during the trial. For example, they might ask the subjects whether they were consciously aware of some stimulus, and how it consciously seemed to them; again, they might ask the subjects whether they were consciously aware of making some decision.

To see how these subjective reports are crucial to this kind of research, compare an analogous investigation conducted with non-verbal but otherwise intelligent mammals, like vervet monkeys, say. You prompt the monkeys in various ways, you get them to perform various tasks, and you check what is going on in their brains at the same time. This research might reveal all kinds of interesting things about monkey cognition, and in particular about the way in which certain causal roles are realized in monkeys. However, it won't tell us anything at all about the monkeys' phenomenal consciousness. Without any first-person reports to go on, it is perfectly consistent with such investigations that monkeys have no phenomenal consciousness at all, or a full phenomenal life just like ours, or anything in between. If we want to find out about the referents of phenomenal concepts, as opposed to merely psychological ones, it seems that we need the subjects to tell us what they are feeling.

We might usefully compare the role of subjects' first-person reports in consciousness research with that of *observation* reports in normal scientific research. When scientists seek to uncover the nature of some natural kind, like water or temperature, say, they will typically begin with some direct observational judgements that certain things are water, certain things are hotter than others, and so on. And then

[4] EEG measures electrical brain waves using electrodes pasted to the scalp. PET scans use radioactive markers in the blood to measure brain activity, while MRI scans achieve the same effect by placing the brain in a powerful magnetic field; these latter scanning techniques use computers to generate striking images of brain activities associated with different mental tasks.

they will seek to construct a theory which will identify further scientifically interesting properties which are common to these observationally identified samples. Similarly with research into phenomenal consciousness. We start with subjects' first-person reports of when they are in pain, seeing an elephant, and so on. And on this basis we aim to develop a theory which will tell us about the material constitution of these states.

Having offered this analogy, let me immediately qualify it. Subjects' phenomenal reports may share the non-inferential directness of ordinary sensory observation, but there are also important differences. For a start, it seems wrong to posit some inner 'phenomenal sense-*organ*' to stand alongside sight, hearing, and so on. The workings of first-person phenomenal judgements was sketched only briefly in Chapter 4, but none of the cases discussed there seems to call for a cerebral mechanism which is *causally sensitive* to conscious experiences. A better model is that first-person phenomenal judgements *incorporate* the experiences they refer to, or re-creations thereof.

Relatedly, the *authority* of subjects' first-person reports in consciousness research is greater than that of ordinary observation reports. In other areas of science, observation reports are defeasible, either on the grounds that conditions of observation are non-standard, or also, on occasion, because later scientific discoveries come to show that previously trusted types of observations are unreliable. By contrast, subjects' first-person reports about their subjective states are standardly immune to these kinds of errors. As I explained in Chapter 4, the correctness of standard first-person judgements simply falls out of the special quotational-indexical structure of phenomenal concepts. If I judge phenomenally of some current state of perceptual classification that *it* is like *this*, there is no real room for me to be wrong; and other kinds of first-person judgements, including recollective judgements, are immune to normal sources of error. Because of this, these subjective reports are not liable to the same kind of correction as ordinary observation reports. (True, when it comes to expressing first-person judgements in a public language, there remains the point that the subjects' *words* can fail to express their phenomenal concepts, and that their *utterances* may be false for that reason; but this does not alter the underlying claim that their non-linguistic judgements are immune to ordinary error.)

However, while I shall generally assume in what follows that subjects' first-person phenomenal reports are immune to normal observational errors, this special authority will play no significant role in my arguments. The more important point is the role that first-person phenomenal reports share with observations in other areas of science. They provide us with an initial sample of cases we can use to get our research off the ground. If we are to identify the material referents of phenomenal concepts, this must be a matter of a posteriori investigation. Subjects' reports can provide us with the database we need to begin this investigation.[5]

7.5 *Consciousness-as-Such*

Before considering in more detail how such investigation might proceed, it will be helpful to introduce a distinction that has not been needed so far. Up to this point, whenever I have given examples of

[5] There is a different kind of 'phenomenal observation' which lacks this significance for phenomenal research. I am thinking here of the kind of case, mentioned in Chapter 4, where you use a phenomenal concept to attribute an experience to another person directly and non-inferentially, as when you judge that someone next to you is in intense pain, say. By any normal standards, such direct third-person phenomenal judgements would seem to count as observational. (We might even postulate a special 'empathetic' perceptual module that non-optionally activates your phenomenal concept of pain in such circumstances.) Even so, I do not take these judgements to have any special methodological significance. This is not because they lack the incorrigibility of first-person phenomenal reports (though they do lack this). Rather, it is because they have no a priori connection with phenomenal concepts. As I observed in Chapter 4, it is an a posteriori matter whether such third-person uses respect the referential content of the relevant phenomenal concepts. As such, these uses will therefore be posterior to empirical research into phenomenal concepts, not a starting-point for such research. This makes intuitive methodological sense: even if an empathetic pain module is automatically activated when monkeys (or humans) display certain publicly observable behaviours, it seems wrong to place the resulting third-person judgements on a par with subjects' self-reports as evidence for the presence of phenomenal pain.

phenomenal concepts, I have always used concepts of specific phenomenal properties, like the concept of feeling pain, of seeing something red, or of hearing middle C. However, these specific concepts are all determinates of the determinable phenomenal concept *consciousness-as-such*. By way of analogy, contrast the determinable, *shape*, with the determinates *square*, *triangular*, *elliptical*. Or again, contrast the determinable *motor car*, with the determinates *Ford*, *a Rover*, *a Rolls-Royce*, and so on. The idea here is simply that of a genus, which then divides into a number of more restrictive species. In this way, then, *consciousness* is a determinable phenomenal concept, whose determinates are more specific phenomenal concepts like *seeing something red*, *feeling pain*, *hearing middle C*, and so on.

Now, much scientific research into phenomenal consciousness is concerned with the material referent of this general concept, consciousness-as-such, rather than with the referents of any more determinate phenomenal concepts. Indeed, a wide range of theories about this general material referent is currently on offer. Thus, to pick a quick sample, consider the identification of consciousness with quantum collapses in cellular micro-tubules (Penrose 1994), or with operations in the global work-space (Baars 1988), or with competition for action control (Shallice 1988), or with representational content (Tye 1995, Dretske 1995), or again, with higher-order thought (Armstrong 1968, Rosenthal 1986, Lycan 1996, Carruthers 2000). These are all materialist theories of what it takes to be conscious at all, not materialist theories of more determinate phenomenal types like seeing something red, feeling a pain, and so on.

In connection with this kind of research, some readers might wonder whether we do in fact possess a phenomenal concept of consciousness-as-such, in addition to phenomenal concepts of determinate conscious states. As we have seen, phenomenal concepts depend on powers of imaginative re-creation or introspective classification. Now, I have argued throughout this book that we can imaginatively re-create and introspectively classify determinate conscious experiences, like seeing something red, say. But do we ever imaginatively re-create *conscious experience* in the abstract, as it were? And do we ever introspectively classify some state simply as *conscious*, as opposed to classifying it as some more determinate conscious state?

I am not sure how to answer these questions. Because of this, I am not confident that there is a phenomenal concept of consciousness-as-such which precisely matches phenomenal concepts of determinate conscious states. On the other hand, it does seem clear that we do have some special way of thinking about consciousness-as-such which is a priori distinct from any material concept of consciousness-as-such. (Given any characterization of some general physical or functional property, it will always seem quite *conceivable*—even if it is not possible—that creatures with that property may not be phenomenally conscious-as-such.)

Perhaps we think about consciousness-as-such phenomenally via *generic* uses of determinate phenomenal concepts. Berkeley held that we cannot imagine triangles-as-such; but he allowed that we can nevertheless prove theorems about triangles-as-such; we imagine some specific triangle, and then ignore its specific features in the proof. Similarly, perhaps we think phenomenally about consciousness-as-such by thinking first about some determinate mode of phenomenal consciousness, and then ignoring its special features in our reasoning.

I shall not pursue this point any further. Let us take it that we do have some kind of phenomenal concept of consciousness-as-such, even if it doesn't work in quite the same way as phenomenal concepts of determinate conscious properties. This will be enough to underpin the enterprise of theorizing about phenomenal consciousness-as-such.

In the rest of this chapter I shall be commenting both on general theories of consciousness-as-such and on specific theories about the material nature of determinate phenomenal concepts. The immediately following sections will focus on determinate phenomenal concepts; I shall turn to consciousness-as-such in sections 7.10–7.15. In many respects the two kinds of theorizing share a similar methodology. In both cases scientists seek to correlate material goings-on with first-person phenomenal reports, hoping thereby to identify some material property as the referent of the phenomenal concept under investigation. But not everything that goes for one will go for the other, and on some points it will be important to treat the two kinds of theorizing separately.

7.6 *Methodological Impotence*

Let me now consider in more detail exactly what kinds of finding we might expect from standard research into phenomenal consciousness. The basic strategy, then, is to take some sample of human subjects, and ask them whether they have some phenomenal property. Simultaneously, we investigate these subjects on a material level, in the hope of identifying some material property which is identical to that phenomenal property.

If the phenomenal property is to be *identical* with some material property, then this material property must be both necessary and sufficient for the phenomenal property. In order for this requirement to be satisfied, the material property needs to be present in all cases where the human subjects report the phenomenal property—otherwise it cannot be necessary. And it needs to be absent in all cases where the human subjects report the absence of the phenomenal property—otherwise it cannot be sufficient. The aim of standard consciousness research is to use these two constraints to pin down unique material referents for phenomenal concepts.

Now, the trouble is that there will inevitably be a *number* of material properties which satisfies these two constraints for any given phenomenal property. The empirical research will of course be able to rule out a large number of candidate material properties, as violating one or the other requirement. Consider the phenomenal property of seeing something red, say. Any material property that is sometimes absent when subjects report seeing something red—activity in some very specific region of the visual cortex, only activated by crimson things, say—is thereby disqualified as *unnecessary* for seeing something red. And any material property that is sometimes present when subjects report that they *aren't* seeing something red—activity anywhere in some larger region of the visual cortex, activated by any colour experience, say—is thereby disqualified as *insufficient* for seeing something red. However, even after we have done all the winnowing out of material properties that can be done by these means, there will still remain a plurality of material properties that might be identified with seeing something red.

Let me illustrate the point with a familiar pair of alternative material properties. Later I shall show that there are various other troublesome alternatives.

Suppose, for the sake of the argument, that we have identified some strictly *physical* property which is present in all and only those human beings who are seeing something red. If this is so, then there will surely be some '*higher*' property which is similarly common and peculiar to just those human beings.[6] Simply abstract away from the details of exactly which molecules are involved, and note what causal organization these molecules ensure. Then the higher property of having this causal organization will also be present in all and only those human beings who are seeing something red.

Yet this higher property will not be identical with the strictly physical property. For there are possible beings who share human higher causal-organizational properties, but not physical properties. We need only consider a 'silicon doppelganger' once more—that is, a being whose cognitive causal structure matches human causal structure, down to a fine level of detail, but is in fact made of silicon-based compounds in place of our carbon-based compounds. So now we face a question: does this doppelganger have our experience of seeing something red when it is confronted with a ripe tomato, or not? Equivalently, should we identify the phenomenal property of seeing something red with the higher property which the doppelganger shares with us, or with the carbon-based property which it lacks?[7]

[6] Recall my terminology from Chapter 1: 'higher' is a generic term, intended to be neutral between physically realized higher-order properties, properties which supervene on physical properties, and properties which are disjunctions of physical disjuncts.

[7] As William Lycan has observed, there will be many 'higher' material alternatives, not just one. Any given human will be materially characterizable at a number of different organizational levels (physiological, neuronal, computational . . .). *Each* of these will present a rival to any strictly physical characterization

At first sight there may seem to be an obvious strategy, in principle if not in practice. Doesn't our difficulty simply call for further dissociative data? We want to decide whether seeing something red is identical to some strictly physical property or to some higher property. So what we seem to need is a creature who has the one material property, but not the other. The conscious state of this creature should then tell us which material property is really identical with the phenomenal experience in question.

In fact, there is only a possibility of dissociation in one direction here. Since physical constitution fixes causal structure, there is no possibility of a creature who has the relevant strictly physical property yet lacks the higher property. But there is the possibility, in principle at least, of dissociation the other way round. This would require a creature whose brain has the right causal structure, but a different physical constitution. A silicon doppelganger would do—or indeed a damaged human in whom the parts of the visual cortex normally involved in seeing something red are replaced by some functionally suitable, but silicon-based, prosthesis.

So the idea is to see whether such a creature still experiences seeing something red. If it does, then seeing something red must be identical with the higher property; if it doesn't, then seeing something red will be identical with the strictly physical property.

But now the limitations of our empirical methodology become apparent. The canonical way of finding out what someone experiences is to take note of their *reports*. Well, it is clear that this creature will *utter the words* 'I am now seeing something red' when

(cf. Lycan 1987). Still, I shall simplify by ignoring this point in most of what follows. A related issue: in presenting the doppelganger problem, I have assumed that *seeing something red* is uniformly physically realized across humans. The problem is then whether to identify the phenomenal property with this uniform realization or some higher property. Given this, it might seem that the problem will go away for phenomenal properties which aren't so uniformly realized, on the grounds that the higher property will then be the only candidate left. But Lycan's point shows that the underlying problem will remain, since there will still be a number of different higher candidates. In view of this, there will be no loss of argumentative generality if I stick to the kind of case where one of the competing material properties happens to be a uniform *physical* realization. (As it happens, I expect that seeing red is uniformly realized in humans. But it is possible that other phenomenal types—sophisticated emotions, perhaps—are not.)

appropriately stimulated. After all, by hypothesis, it is causally structured just like a normal human being. So the state produced in it by red tomatoes will be linked up to its language processors and motor cortex just as the corresponding state is linked up in humans, which ensures that the creature will make just the same verbal reports. So at first pass this would seem to argue for the identification of seeing something red with the higher property that this creature shares with humans. The creature has this higher property, lacks the normal human physical property, and says 'I am now seeing something red'.

But of course this test is indecisive. For the creature would clearly make just the same report *even if* the phenomenal experience of seeing something red were identical with the physical property it lacks, and not with the higher property it has.

The trouble is that we do not know what the phrase 'seeing something red' refers to in this creature's mouth. If we could be sure that it referred to the normal human experience of seeing something red, then the creature's report that it is seeing something red would indeed show that the experience goes with the higher and not the strictly physical, property. But we can't be sure that seeing something red in the creature's mouth refers to seeing something red. Rather, it refers to whichever kind of experience it has, and we don't yet know what that is.

The evidence provided by the creature can be construed in two ways. *If* experiences go with higher properties, then the creature will share our experience, and its 'seeing something red' will refer definitely to this shared experience. On the other hand, if experiences go with strictly *physical* properties, then the creature won't share our experience of seeing something red, and its 'seeing something red' will refer definitely to something other than *our* experience of seeing something red.

Thus the upshot is that the empirical methodology I have outlined is impotent, even in principle, to identify precise referents for determinate phenomenal concepts. The methodology can show us that an experience like seeing something red is precisely correlated with some strictly physical property in humans. But it will also show that this experience is precisely correlated with various higher properties. Given this, we are stuck. It is no good finding a test creature in whom the higher properties are differently realized, and asking it whether it still has the experience. For it will inevitably say 'Yes' whenever it has the higher properties, but we won't be able to assign a definite meaning to this.

7.7 *Further Alternatives*

Readers of Wittgensteinian inclinations may feel that this supposed ineffability in the phenomenal utterances of differently constituted creatures simply shows that there is something very wrong with my overall account of phenomenal concepts. After all, they may say, it is scarcely surprising that we should conclude that we are unable to interpret certain mouthings, once we suppose that those mouthings express concepts constituted out of essentially private experiences. The problem, on this diagnosis, would derive from the initial supposition that phenomenal concepts can refer in their own right, independently of any a priori tie to publicly accessible material occurrences.

I reject this diagnosis. I argued in Chapter 4 that the 'privacy' of phenomenal concepts, such as it is, is no barrier to their having semantic powers in their own right. The problem we have now run into does not show that there is anything wrong with that argument, or that differently constituted creatures cannot think meaningfully with purely phenomenal concepts, just as we can. The reason we have difficulty understanding the phenomenal utterances of such creatures isn't that these utterances are somehow semantically empty. Rather, the difficulty stems from the fact that their phenomenal concepts, like ours, are *vague*.

However, rather than pursue this point now, let me leave it until the end of section 7.9, after I have developed and defended my general thesis that the limitations of consciousness research stem from the vagueness of phenomenal concepts. It will then be easy to see that the semantic indeterminacy of the phenomenal reports of differently constituted creatures is simply a corollary of the general vagueness of phenomenal concepts.

Before proceeding to this general issue of vagueness, it will be helpful briefly to show that there are other versions of the problem raised in the last section. I there showed that our empirical methodology is impotent to decide between strictly physical properties and higher material properties as the material referents of phenomenal concepts. It is easy to see that there are a number of further such choices between which our methodology is similarly impotent to decide.

Thus suppose, for the sake of the argument, that empirical research indicates some *representational* material property as a possible candidate for the material referent of some phenomenal concept. For example, suppose that empirical research shows that subjects are disposed phenomenally to report that they are seeing an elephant when and only when they embody some characteristic cognitive representation of an elephant.

Now, the relationship between phenomenal consciousness and representation raises many issues, some of which will be mentioned in section 7.15 below, when I consider representational theories of consciousness-as-such. But for the moment let me simply specify that I am here thinking of representation as a *material* matter: so when I say that someone embodies some characteristic cognitive representation I should thus to be understood as conveying that they satisfy some material concept of representation, of the kind that might be found in a causal or teleosemantic account of representation.[8]

The difficulty I want now to address arises because representational material properties can be individuated *broadly* or *narrowly*. It is now common, at least among theorists working on representation itself, in abstraction from any connection with consciousness, to argue that representational properties are typically *broad*, in the sense that their possession is fixed, not solely by matters inside the subject's skin, but also by extracranial relations to the representation's subject-matter. On this view, to have a cognitive representation of

[8] That representation can be thought about materially leaves it open that it can also be thought about phenomenally. Indeed, I find it highly plausible that certain phenomenal concepts present their referents as representational. I shall return to this point in section 7.15.

elephants, say, requires that you bear certain causal or historical relationships to elephants. (Sometimes this broadness is defended purely on intuitive grounds. But it is also a corollary of most reductive accounts of representation, including standard teleosemantic and causal accounts.)

The point is made graphic by the standard thought-experiments. Imagine a being who is an exact physical duplicate of me, but who has a non-standard physical or social environment. (Cf. section 1.5.) Despite the intrinsic identity, it strikes many people as intuitively wrong to suppose that this being can represent features which its non-standard environment prevents it from interacting with. It seems intuitively wrong, for example, to hold that a physical duplicate living on a physically similar distant planet can represent *Marilyn Monroe*, say, or indeed represent *elephants*. (Moreover, these intuitions are supported by theories which hold that representation derives from causal or historical relations, since the duplicate will lack the appropriate relations.)

If you combine a broad account of representation with a representational account of some phenomenal property P, then the implication is that environmentally different duplicates will lack P, despite their intracranial identity with people who have it. Some philosophers are prepared to bite this bullet (cf. Dretske 1995, Tye 2000). But rather more regard this conclusion as unpalatable: surely, they say, what it's like phenomenally must be fixed by what's inside your skin, and not by things outside you. This latter intuition does not necessarily mean giving up a representational account of P altogether. For there remains the option of factoring out a 'narrow' representational property from the initial broad one, and identifying P with that instead. On this approach, broad representational properties are viewed as decomposable into two factors: a narrow factor, which is fixed solely by how things are inside the head of elephant-representers, say, and so will be shared by duplicates, and an external addendum, which involves relationships to actual elephants, and which duplicates will lack. If you are attracted to a representational account of P, yet think that my duplicate must feel just like I do, then you need to identify P with some narrow representational property, rather than a broad one composed of that narrow property plus the external addendum.

So representationalism about phenomenal properties like seeing an elephant can be developed in two different ways. Some representationalists are prepared to identify this kind of phenomenal property with a broad representational property, while others will settle for identifying it with a narrow one. This thus gives us another case where we have two competing candidates for the material referent of a phenomenal property. Moreover, it is another case where our empirical methodology is impotent to decide between the competing candidates.

To see this, consider the kind of test case which would decide between such a narrow and a broad candidate. We need a creature, like a human duplicate on a distant planet, who has the narrow property but lacks the external addendum. If this creature has the phenomenal property at issue, then we can rule out the broad candidate, for the external addendum will have proved not to be necessary for the phenomenal property. Conversely, if this creature lacks the phenomenal property, we can rule out the narrow candidate, for it will have proved insufficient.

The trouble, as before, is that the canonical way of finding out what a creature experiences is to take note of its *reports*. But we already know what *words* the duplicate in the narrow state will utter, independently of whether it shares our phenomenal experience of seeing an elephant. Since it differs from us only in its external relations to elephants, and not in terms of its current constitution, its mouth will move just as our mouths move, and it will utter the words 'I am consciously seeing an elephant'.

But of course this doesn't establish that the duplicate shares our phenomenal experience of seeing an elephant. For, just as with the silicon doppelganger in the last section, we cannot determine what the words 'seeing an elephant' refer to in the duplicate's mouth. If they refer to the normal human experience of seeing an elephant, then the duplicate's report would indeed show that seeing elephants goes with narrow, and not broad, representational properties. But nothing in the data rules out the alternative possibility, that the duplicate's words refer to something else, and that the broad representational property is indeed necessary for *our* experience of seeing red. To evaluate the significance of the duplicate's words, we need to know what they refer to, and we have no independent way of determining this.[9]

Let me conclude this section by briefly drawing attention to one further kind of competition for the material nature of phenomenal properties. Consider the view that a mental state is only conscious if the subject makes a Higher-Order judgement about that state (where by 'Higher-Order' I mean a self-referential judgement to the effect that the subject is in some mental state[10]). Now, there are different versions of such Higher-Order theories of consciousness. Some of the issues they raise will be discussed later, when I consider Higher-Order theories of consciousness-as-such in sections 7.11–7.13. But it is easy enough to see that the possibility of such theories creates alternatives between which the standard methodology will have trouble deciding.

For suppose, in line with the Higher-Order approach, we are offered the hypothesis that some phenomenal property P is identical

[9] Even if our empirical methodology can't decide between narrow representation and standard versions of broad representation as the material basis of perceptual experience, it does, I think, rule out one radical possible species of externalism about perceptual experience. I am thinking here of a strongly externalist or 'disjunctivist' view, according to which the nature of your phenomenal experience is fixed by your current particular context (rather than your historical or normal context), in such a way that hallucinators have different phenomenal properties from veridical perceivers. Whatever the merits of distinguishing veridical perception from hallucination for explanatory or epistemological purposes, there seems no way for consciousness research, as I am conceiving it, to warrant viewing them as different *phenomenal* types. Suppose we posited that the current presence of a real elephant is necessary for some phenomenal state ('seeing an elephant', let us call it). The trouble is then that hallucinating subjects will also first-personally report themselves to be in this phenomenal state. And this means, according to the standard methodology, that a real elephant can't be necessary for the phenomenal state after all. Our database will include cases where ordinary humans report first-personally 'I am *seeing an elephant*', even though no real elephant is present.

[10] Note that this sense of 'Higher-Order' (which applies to mental judgements *about* mental states) is quite distinct from the metaphysical sense of 'higher-order' (which applies to a state of having-some-property-which-satisfies-some-constraint). States which are 'higher-order' in this latter metaphysical sense need not represent anything. I shall use initial capitals ('Higher-Order') to distinguish the representational notion.

(a) with some underlying material correlate M *plus* a Higher-Order judgement that P is present. Now consider the alternative thesis (b) that P is identical with M alone. Skipping over various complexities, to be addressed in sections 7.11–7.13, we can immediately see that the standard methodology runs into trouble with the choice between (a) and (b). For this methodology advises us to seek some material property characteristic of cases where subjects phenomenally report P. But any such reported cases which have M will also have M-*plus*-a-Higher-Order-judgement-that-P-is-present. (How can subjects report their states, if they don't know *about* them?) Moreover, any cases with M-*plus*-a-Higher-Order-judgement-that-P-is-present will also obviously have M. So any cases which support (a) will also support (b), and vice versa. The standard methodology thus offers no obvious way of prising apart (a) and (b) as specifications of the material nature of P.[11]

7.8 *Vague Phenomenal Concepts*

As I said earlier, Ned Block has argued that the difficulties raised in the last two sections are the Achilles' heel of inflationist materialism. Inflationist materialism, remember, is distinguished from earlier deflationist approaches to consciousness by its recognition of phenomenal concepts which are a priori distinct from any material concepts. By contrast, deflationists do not acknowledge phenomenal concepts, and hold that everyday thought conceives of conscious states solely in psychological terms which refer via descriptions of causal roles.

Now, I have argued throughout this book that deflationism is unable to answer the standard objections to materialism, and that any satisfactory materialism must therefore be inflationist. But it is noteworthy that, whatever its other drawbacks, the deflationist alternative does not face the same methodological impasse that we

[11] If any readers are thinking that (a) may be shown to be superior to (b) by *negative* cases, where subjects with M but no Higher-Order awareness of P report that they *don't* have P, I ask them to wait until section 7.13, where I have a proper look at this can of worms.

have reached in the last two sections. Inflationists seem to run into an epistemological barrier: the way they conceive of consciousness seems to condemn them to perpetual ignorance about when it is present in non-human beings. But no such consequence is forced upon deflationists. From their point of view, empirical research is an uncomplicated matter: observation shows us when and where psychological roles are satisfied, and more detailed physiological investigation then uncovers the different physical states which play these roles in animals, androids, and any further strangers who turn out to satisfy the roles.

I do not think that this comparison reflects as badly on inflationism as Block suggests. There would indeed be something puzzling if inflationism implied that there were definite facts about consciousness that no amount of empirical research could possibly uncover. If there were such definite but inaccessible facts, then surely there ought to be some explanation of why we can't find out about them. But it is not clear that inflationism can offer any such explanation. It is not as if the facts of consciousness are too far away, or too small for our instruments, or anything like that.

However, I do not think that there are definite facts about consciousness that lie beyond our epistemological grasp. Rather, I think that the reason we can't answer certain questions about consciousness is simply that our phenomenal concepts are *vague*. There is nothing in the workings of phenomenal concepts like *seeing something red*, or *being in pain* to fix whether or not silicon doppelgangers or other-worldly duplicates satisfy these concepts. It is not that there is some fact of the matter here, which we lack access to. Rather, *all* the facts will fail to fix an answer, for our concepts are not sharp enough to determine whether doppelgangers and duplicates fall within their boundaries or not. Even God, who knows everything, will not know whether these beings satisfy our concepts of *seeing something red* or *being in pain*.[12]

This response might seem cheap. The suggestion that phenomenal concepts are vague does offer an answer to Block's challenge. But, if

[12] I first argued that concepts of conscious states are vague in Papineau 1993*a*, 1993*b*, and 1995. But I did not there distinguish clearly between vagueness in phenomenal concepts and vagueness in psychological ones.

this is all there is to be said in its favour, it is surely *ad hoc*. After all, the intuitively more natural view is surely that either doppelgangers and duplicates will have the relevant experiences, or they won't. In the absence of independent arguments for vagueness, it would seem that Block is justified in his claim that inflationists have saddled themselves with an inexplicable barrier to discovery.

Fortunately, I think there is ample independent reason to think that phenomenal concepts are vague. If we refer back to our earlier analysis of phenomenal concepts, we can see that there is nothing in their semantic workings that could possibly ensure that they refer to one rather than another of the material properties which are characteristically present when normal humans report that they are phenomenally seeing something red or are in pain.

I argued earlier, in Chapter 4, that the referential power of phenomenal concepts is at bottom a causal or teleosemantic matter. Phenomenal concepts refer in virtue of the characteristic causes or biological functions of the judgements they enter into. However, any causal or teleosemantic account will leave it indeterminate exactly which of the correlated material candidates any given phenomenal concept refers to. For all the correlated material candidates will figure equivalently in the characteristic causes[13] or biological functions of the relevant phenomenal judgements, and so causal or teleosemantic considerations will fail to pick out one material candidate rather than another as the referent.

We can think of phenomenal concepts as tools which enable us to track facts involving human experiences. In effect, phenomenal concepts enable us to categorize ourselves and other humans as undergoing certain experiences. But experiences are material states.

[13] Could a *broad* mental state be a characteristic cause of the relevant phenomenal judgements? Aren't my phenomenal judgements caused by matters inside my skin, not outside? Well, this may be true of first-person phenomenal judgements, but it is obviously not true of third-person ones. In any case, if there is a difficulty here, I take it to reflect badly on causal theories of content, and to favour teleosemantic ones. If causal theories are condemned to read concepts as always referring to their most proximal characteristic causes, then so much the worse for causal theories: they won't be able to read the concept of *sunburn* as referring to burning caused by the sun, for example, or the concept of *fossil* as referring to mineralized relics of past animals.

So phenomenal concepts serve to track facts involving material properties. But which material properties precisely? There are various different candidate material properties, each of which would serve to make the same categorizations among human beings. Given this, there is no reason to suppose that phenomenal concepts serve to track one of these material properties rather than another. Since the same categorization of human beings would result in any case, we can conclude that phenomenal concepts refer indeterminately to any of those material properties.

In effect, phenomenal concepts are crude tools. They have no theoretical articulation which might tie them to strictly physical properties rather than higher ones, or to narrow properties rather than broad ones, or to other targets among competing material referents. They do their job adequately as long as they enable us to respond to the packages of co-occurring material properties associated with experiences. Since these packages never come undone in normal human beings, nothing decides which of the material properties they contain are the referents of phenomenal concepts.

7.9 *Vagueness Defended*

It may seem very odd to hold that a phenomenal term like 'seeing something red' is vague, and that there is therefore no fact of the matter of whether a silicon doppelganger looking at a ripe tomato is seeing something red or not. Surely, you may feel, either it is visually *like this* for the doppelganger, or it is not. What could be more clearly a matter of fact than that?

In this section I want to consider this worry, and show that my thesis of vagueness is not as odd as it might seem at first (though I do not deny that it is still pretty odd). It will be helpful to focus for the moment on the determinate phenomenal property 'seeing something red' and its putative vagueness as applied to silicon doppelgangers.

Let me be clear about the precise point at issue. My claim is not that it is vague how it is for the doppelganger. The doppelganger's experience will feel as it does, and there is no need to suppose that this in itself is less than definite, that there is somehow some fuzziness in the doppelganger's experience itself. Rather, my claim is that our phenomenal term 'seeing something red', the one whose exercise involves instances or reactivations of our own red experiences, is not well focused enough for it to be determinate whether or not the doppelganger's experience falls under it. This term works well enough in discriminating normal human beings one from another in respect of whether they are seeing something red. But when we seek to apply the term beyond the cases where it normally works, it issues no definite answer.

In the normal human case, our phenomenal term 'seeing something red' distinguishes effectively between those who have both some physical property and a higher property which is fixed by that physical property, and those who have neither of these properties. But now we are asking the term to decide what we should say about a being who has the higher property but not the physical property. There is no reason to suppose that there is anything in the workings of the term to decide this question.

Again, we needn't suppose that there is anything less than definite in the doppelganger's experience itself. For the doppelganger, the experience will feel as it does. The question is rather whether an experience which feels like *that* is sufficiently similar to the normal human experience of seeing something red to fall under our term 'seeing something red'. (Similarly, it might be indeterminate whether some experience induced by a hallucinogenic drug, or produced by weird lighting, or deriving from the synaesthetic appreciation of a sound, should count as 'seeing something red'.)

Doubters are likely to remain unconvinced. They may feel that either the doppelganger's experience is *exactly like this* colour experience (and here they imagine a red colour experience), or it is not. Surely this must admit of a definite answer (for God at least, even if the answer is not available to us).

But consider this analogy. Surely my friend's head of hair is exactly like mine in respect of being bald, or it is not. Well, if my friend and I are strictly physically identical, then surely we are alike in baldness. For it seems clear that, whatever 'baldness' may refer to, it must refer to some property that is fixed by strictly physical constitution. Two people can't differ in being bald without differing physically.

Similarly, we can take it, two beings that are exactly identical physically must indeed be alike in whether they are 'seeing something red', and for the same reason. Whatever phenomenal concepts refer to, they must at least refer to something that is fixed by strictly physical constitution, as was shown by the arguments in Chapter 1.

But now consider a friend whose hair is similar to mine in some ways, but not others. Maybe he has the same number of hairs, in the same places, but his hair is of a different texture or colour. Or maybe he has far fewer hairs, but they are somehow thicker than mine. Now ask whether my friend's head of hair must be exactly like mine in respect of being bald, or not. It is not clear. 'Bald' is a vague term, and different refinements of the term may issue in different verdicts on whether my friend is exactly as bald as I am. As we use the term, there need be no fact of the matter as to whether we are exactly equally bald.

Thus too, I say, with the doppelganger's visual experience. A being who is exactly like me physically will indeed be just like me in respect of seeing something red. But there need be no determinate answer for a being who shares some of my material properties but not others. Different ways of refining the term 'seeing something red' will issue in different verdicts. So our actual unrefined use of the term fails to decide whether the doppelganger is just like me in seeing something red, or not.

I can now deal with a query left hanging at the beginning of section 7.7: why can't we assign a definite meaning to the silicon doppelganger's phenomenal reports? The Wittgensteinian suspicion was that this is an unsurprising upshot of my misplaced enthusiasm for 'private languages'. However, we can now see that this isn't the reason at all. Rather, the point is simply that the doppelganger's phenomenal terms are vague, just as ours are, and for the same reason. Just as it is indefinite whether the phenomenal concept that *we* express by 'seeing something red' applies to the doppelganger's ripe tomato experience, so it is indefinite whether the phenomenal concept that the doppelganger so expresses applies to *our* ripe tomato experience.

Let us take it that the doppelganger is like a human in all respects, historical and contextual, bar its basic physical constitution. Then, just like us, it will have phenomenal concepts, whose exercises incorporate its mental states or re-creations thereof, and which thereby refer to those selfsame states. In particular, such a phenomenal concept will be expressed by the doppelganger's words 'I am now seeing something red'. This phenomenal concept will pick out those doppelgangers who are looking at red things. These doppelgangers will share a higher material property with humans, and this will be realized by a silicate property which they do not so share. Now, does the doppelganger's concept here refer to the material property or the silicate property? This will decide whether the concept applies to other beings, like humans, who have the material property but not the silicate property. But there is no fact of the matter here. Since the doppelganger is effectively a mirror image of us humans, all the considerations that apply in our case will also argue that the doppelganger's corresponding concept fails to decide between the two properties. And this finally, rather than any Wittgensteinian difficulty, is why we can't assign a definite meaning to the doppelganger's utterances.

7.10 *Theories of Consciousness-as-Such*

So far I have focused on the thesis that *determinate* phenomenal concepts are vague—there need be no fact of the matter as to whether certain creatures are seeing something red, or feeling pain, or seeing an elephant. But what about the determinable, consciousness-as-such? Is this vague too? This would seem an even odder claim. It is one thing to argue that it is vague whether octopuses count as being in pain, or silicon humanoids as seeing something red. Perhaps here our determinate concepts do indeed fail to draw sharp lines. But it is another thing to argue that it can be vague whether such creatures are conscious at all. Even if you have been persuaded by my arguments so far, you may be likely to gibe at this further claim. For surely, it seems, there must be a fact of the matter whether it is *like anything at all* for such creatures. Maybe there is unclarity about how exactly to classify specific states of consciousness in alien creatures. But it can't be unclear whether they have any such states to start with. Either there is some spark of consciousness present, or there isn't.

Despite the plausibility of this line of thought, I want to argue that even the determinable concept consciousness-as-such is vague. There need be no fact of the matter about whether or not certain creatures are phenomenally conscious. The problem does not stop with specific modes of consciousness. Even the general concept consciousness-as-such fails to draw a sharp line through nature.

I realize that this claim will strike many readers as hard to swallow. Even so, I hope to to render it plausible. Let me start by drawing attention to one obvious circumstance which adds to the apparent oddness of my claim. This is the strong dualist intuition that phenomenal properties are distinct from any material properties. If you accept this dualist intuition, then you will think that it must be determinate whether phenomenal consciousness is present or not. If consciousness is an extra inner light, so to speak, distinct from any material properties, then there must always be a definite fact of the matter whether this light is switched on, however dimly, even in unfamiliar cases.

I agree that the imputation of definiteness would follow, *if* dualism were true. But since I don't accept dualism, I simply regard it as a yet further illustration of the way in which the dualist intuition of distinctness distorts our thinking about phenomenal consciousness. Once we free ourselves from this intuition, then perhaps we will not feel so sure that questions about phenomenal consciousness must admit definite answers. If reality contains nothing but various species of material properties, and no distinct phenomenal properties, then perhaps it will not seem so surprising that our concept of phenomenal consciousness should fail to cut nature at a sharp seam.

It is my strong suspicion that much empirical research into consciousness is motivated by dualist intuitions. I have in mind here not only those researchers who explicitly endorse dualism, but also many of those who deny it. Thus theorists who begin by explicitly disavowing any inclinations towards dualism will often betray themselves soon afterwards, and slip into the familiar talk of brain processes as 'generating' consciousness, or 'causing' it, or 'giving rise to' it, or 'being correlated with' it,[14] or any of the other phrases which trip so easily off the tongue, but which only make real sense if conscious properties are distinct from material properties.

In so far as this is what drives the current boom in consciousness research, then I think the boom is quite mismotivated. There is no extra stuff, over and above material stuff, to distinguish beings with phenomenal consciousness from those without. So there is no question of finding out about any such extra stuff.

I shall not comment on dualist thinking any further. I take dualism to have been amply discredited by the arguments rehearsed earlier in this book. Rather, my focus from now on will be on the question of what *material* property, if any, our general phenomenal concept of consciousness-as-such refers to. And my answer, as with more determinate phenomenal concepts, will be that there are a number of competing candidates for the material nature of phenomenal consciousness, and no fact of the matter as to which of these candidates our phenomenal concept of consciousness-as-such really latches on to.

7.11 *Actualist HOT Theories*

It will be helpful at this stage to consider one particular family of proposals for a material reduction of consciousness-as-such: namely, Higher-Order proposals, of the sort alluded to at the end of section 7.7. (Cf. Armstrong 1968, Dennett 1978*a*, Rosenthal 1986, Lycan 1996, Carruthers 2000.) Apart from being of interest in their own right, 'HOT'[15] theories of this kind also raise a number of crucial methodological issues.

Let me begin with the most straightforward kind of HOT theory:

[14] Cf. the contemporary use of the catch-phrase 'the neural *correlates* of consciousness', even among scientists who would be most distressed to be told they have dualist inclinations.

[15] 'HOT' for 'Higher-Order Thought'. In fact, not all Higher-Order theorists are happy to characterize the relevant Higher-Order judgements as *thoughts*. But it will be convenient to use the acronym to stand for all Higher-Order accounts of consciousness.

the 'actualist' view that a mental state is conscious if and only if it is the subject of some *actual* Higher-Order judgement. The general idea is that a state is conscious if the subject is 'aware' of it, where this is understood as a matter of the subject forming some actual Higher-Order judgement about it. Later I shall consider a contrasting style of 'dispositional' HOT theory (cf. Carruthers 2000).

I shall assume that the Higher-Order judgements at issue here are first-person phenomenal judgements made using phenomenal concepts, of the kind discussed at length throughout this book. In this respect I shall be going beyond existing advocates of HOT theories, given that they do not invoke the specific analysis of phenomenal judgements that I have developed here. Still, there seems nothing in my analysis of phenomenal judgements to render them unfit for a role in HOT theories; indeed, they seem just the kind of Higher-Order judgements that HOT theories need.[16]

On a first quick reading, the Actualist HOT view looks as if it cannot but receive strong support from our empirical methodology for studying consciousness. Whenever subjects phenomenally report themselves to be conscious-as-such, they will therewith have made a Higher-Order judgement about some phenomenal state, and whenever they phenomenally deny that they are conscious-as-such, they will not have made any such Higher-Order judgement. The presence of a Higher-Order judgement would thus seem to correlate perfectly with subjects' reports about consciousness-as-such. What better evidence could there be that the essential characteristic of phenomenal consciousness-as-such is the presence of a Higher-Order judgement?

But once we probe a little bit deeper, things prove less straightforward. Despite first appearances, Actualist HOT theories do not in fact enjoy any such perfect fit with the empirical methodology. They face an awkward problem in relation to Higher-Order *memory* judgements. To find a good fit with the empirical methodology, we will have to wait until section 7.13 and

[16] In particular, in so far as first-person phenomenal judgements are immune to error (cf. section 4.12 above), a focus on phenomenal judgements promises to free HOT theories from awkward criticisms arising from the possibility of Higher-Order *mis*judgements about phenomenal states (cf. Neander 1998).

'dispositional' HOT theories, which avoid this problem about memory judgements.

To see the problem, consider this case: I see a red pillar-box, form no Higher-Order phenomenal judgement about this at the time, but then later imaginatively recall the experience of seeing something red. This certainly seems initially possible. Moreover, in such cases subjects will presumably later report, on the basis of their later memory judgement, 'Yes, I consciously saw something red earlier'. So the standard methodology will count the earlier experience as conscious: after all, the subject has issued a phenomenal report to this effect, and so the experience will go into the database of cases we use to investigate the material referent of our phenomenal concept of consciousness-as-such.

On the other hand, it is not at all clear that *Actualist HOT theorists* will want to count this earlier experience as conscious. If no introspective Higher-Order phenomenal judgement was made at the time of the experience, then on their view the status of that experience as conscious will presumably have to depend on the occurrence of the later Higher-Order memory judgement. But this seems silly. How can an earlier state be rendered conscious by some later act of memory? What if the act of memory hadn't occurred? Then presumably the earlier state wouldn't have counted as conscious. But surely the status of some state as conscious must be fixed by how things are when it occurs, not by whether or not something happens later.

There are a couple of ways in which Actualist HOT theorists might avoid this awkward backwards causation of conscious status. One possibility would be to deny that the earlier state does qualify as conscious, since no Higher-Order judgement was present at the earlier time. This has a kind of cogency. But it seems quite *ad hoc* in relation to the standard methodology of consciousness research. This methodology regards memory and introspection as on a par as sources of information about phenomenal consciousness. Indeed, memory is rather more important than concurrent introspective judgements in many psychological experiments. Thus, unprimed subjects are subject to some manipulation, and then *afterwards* they are asked, 'What, if anything, did you experience then?' To deny the reliability of these reports would undermine any amount of apparently sound research. Of course, the reliability of phenomenal memories should not be taken for granted: there are well-known confounding effects, such as the tendency to confabulation (cf. Nisbett and Wilson 1977, Nisbett and Ross 1980, Wilson *et al.* 1981). But we can recognize this danger without dismissing *all* phenomenal memories of non-introspected experiences. It seems quite unmotivated to hold that *no* positive memory judgements can ever be accurate about the phenomenal status of non-introspected earlier experiences, even when there are no independent reasons to distrust these memories, simply because Actualist HOT theories would not otherwise make sense. This looks like the theoretical tail wagging the methodological dog.

The other way for Actualist HOT theorists to respond to the problem of experiences which are not introspected but are later remembered would be to argue that there aren't in fact any such cases, because we humans don't in fact ever phenomenally remember anything we didn't introspect at the time. The idea here would be that some kind of priming by current introspection is an empirically necessary condition for the later phenomenal re-creation of an experience—we can't phenomenally bring an experience back unless earlier introspection has served to 'put it in the archives'. If this were right, then any experience remembered as conscious would have been introspected earlier, and Actualist HOT theorists would no longer have to worry about later memories absurdly bestowing consciousness on earlier experiences which weren't then introspected.

The trouble with this line is that it just isn't plausible that earlier introspection is empirically necessary for later phenomenal memory. I can surely walk down the street watching children play, but not thinking *about* my visual experiences, and then later recall imaginatively how it looked to me. If the current suggestion were right, it would require an absurd amount of introspective activity for us to be able to imaginatively remember all the things we can so remember. Or, to put the point the other way around, we would be able imaginatively to remember much less than we can, given how rarely we introspect, if the current suggestion were right.

7.12 *Attention*

The implausibility of this last suggestion—that we can't phenomenally remember things we didn't introspect at the time—can be obscured by its conflation with a different and far more plausible thesis: namely, the thesis that earlier *attention* is necessary for later phenomenal memory.

By 'attention' here I mean the kind of state I called 'perceiving as' in Chapter 4. In that chapter I conceived of 'perceiving as' in terms of the 'intensification' or 'highlighting' of experiences, and I suggested that this might arise because incoming stimulations match or resonate with some stored perceptual pattern, where this stored pattern will standardly derive from previous sensory experience. (Thus, in general, it would be impossible to perceive something *as* Ø unless you had already been exposed to Øs.[17])

Now, as I told the story in Chapter 4, such attention, or 'perceiving as', is a necessary condition for the *introspective* use of phenomenal concepts. In order effectively to construct an introspective phenomenal term of the form 'the experience: ---', you need to highlight some aspect of your current overall manifold of experience, in order for it to be determinate which such aspect your phenomenal term refers to.

The question presently at issue, however, is whether such attention is necessary in order for an experience to be referred to *later* by a re-creative memory use of a phenomenal concept. It is not obvious that it is. Even if I don't highlight a given aspect of my overall experience at the time (and so couldn't then have formed an introspective phenomenal concept), why shouldn't I later be able to re-create that aspect in perceptual imagination, and then use this re-creation to fill the gap in the phenomenal construction 'the experience: ---'? This would then give me the conceptual wherewithal to form a recollective phenomenal judgement about that earlier experience.

[17] I realize that analogous attentional processes occur at many levels in the brain, some far removed from the conscious realm. Let me simply specify that by 'attention' I here mean those relatively sophisticated processes which allow the formation of introspective phenomenal concepts in humans.

However, even if nothing rules this out a priori, there is empirical evidence which suggests that it is not in fact psychologically possible. Experimental subjects who are induced to attend to one thing (to a given spot in their visual field, say) standardly report that they have no phenomenal memory of any items which we might otherwise have expected them to remember perceiving (such as the appearance of some coloured shape near to, but outside, their focus of attention). (Cf. Mack and Rock 1998.)

In what follows, I am happy to take it, then, that it is impossible phenomenally to remember experiences which did not involve *attention* at the time. The point I want to stress, however, is that this does *not* imply that it is impossible phenomenally to remember experiences which you did not phenomenally *introspect* at the time. Attention in itself is something less than introspection. So even if all remembered experiences were attended to at the time, this doesn't mean that they were then introspected.

The point is that an experience can be highlighted by attention without your forming any introspective judgement *about* that experience (such as 'I am now having the experience: ---'). Attention is in the first instance focused on the world, not introspectively on experiences. I can attend to the children playing —see them *as* children playing—without thinking about my visual state.

According to my overall story, then, attention is a necessary condition for introspective judgements—you can't introspect without attention—but it is not sufficient—you can attend without forming any introspective judgement *about* your experience. So the fact that you need to attend to remember phenomenally does not imply that you need to introspect to remember phenomenally.

So, once we distinguish carefully between attention and introspection (as HOT theorists are not always careful to do—cf. Lycan 1996: ch. 2), we can confirm that prior phenomenal introspection is not necessary for later phenomenal memory. The empirical data do offer some support for the claim that *attention* may be necessary for such later memory. But they do nothing to support the different, and antecedently implausible, claim that I can't phenomenally remember an experience (seeing the children playing) unless I was thinking *about that experience* at the time. And this, to return to the original point, is why Actualist HOT theories can't avert the threat of the backwards causation of consciousness by denying that we ever phenomenally remember experiences we didn't introspect earlier. There may be no later phenomenal memories of earlier *unattended* experiences, but there are surely plenty of later memories of earlier *unintrospected* experiences.

7.13 *The Dispositional HOT Theory*

Let me now turn to a rather different kind of 'HOT' theory. This is the 'dispositional' HOT thesis that a state is conscious just in case it *could* have been the subject of an introspective Higher-Order judgement, even if it wasn't actually so subject (Dennett 1978*a*, Carruthers 2000). This version enjoys the general advantages of the HOT approach, while avoiding the backwards causation of consciousness that discredits Actualist HOT theories. Let me first show how this dispositional theory works. I shall then turn to some methodological issues it raises.

The idea behind the dispositional approach is that a state doesn't actually have to be the subject of a Higher-Order phenomenal judgement to count as conscious. It is enough that it would have been the subject of such a judgement, had the thinker addressed the issue at the time. So a particular mental state could be conscious, even if it was not *actually* introspected phenomenally, provided it is the kind of state that can be so introspected. The idea, then, is that a human mental state qualifies as dispositionally Higher-Order judgeable just in case the subject would have applied a phenomenal concept to it if the question of its phenomenal nature had been raised at the time.

Note immediately how this Dispositional HOT theory avoids the difficulty that faced Actualist HOT theories. The difficulty, recall, arose with earlier states which were not introspected at the time, but were later reported as conscious by phenomenal memories. This is no longer a difficulty, since any state which is so remembered phenomenally will be one which the subject could have introspected phenomenally at the time. If a normal human being can remember an experience phenomenally, it must have been highlighted by experience earlier (cf. the last section), and the subject would then have been capable of forming an introspective phenomenal concept from that highlighted experience (cf. section 2.9). So the subject would have classified it introspectively under this concept when it occurred, had the subject then considered the matter. This means that any phenomenally remembered state always was conscious, according to the Dispositional HOT theory. The subject *could* have formed an introspective Higher-Order judgement about the state earlier, even if the subject didn't in fact do so. This removes the danger of such states being retrospectively rendered conscious by later memories.

Now that we have the Dispositional HOT theory in our sights, we can see that this theory is guaranteed to provide a strong fit with the empirical methodology for studying consciousness. To show this, let me consider in turn positive subjects' reports ('I am/was consciously Øing') and negative ones ('I am/was *not* consciously Øing').

If a subject issues a positive report, then clearly the relevant state was dispositionally Higher-Order-judgeable. For, if there is a *current* introspective report ('I am Øing'), then the state is actually Higher-Order-judged, and so *a fortiori* dispositionally Higher-Order-judgeable. And if there is a *later* phenomenal memory report ('I was Øing'), then it follows that the subject would have introspectively classified the state earlier, had the question arisen, for the reasons just given, in the paragraph before last. So the property of dispositional Higher-Order judgeability is *present* whenever there is a *positive* subject's report on consciousness.

Is the property of dispositional Higher-Order judgeability always *absent* when subjects issue *negative* reports on consciousness? This is a bit more tricky. Negative *introspective* reports are all right: a negative current introspective report immediately shows that the subject didn't Higher-Order-judge the state to be conscious, when the question did arise. But memory reports raise problems again. If a subject says later that a certain earlier state *wasn't* conscious, then presumably it follows that either the subject is never able to think phenomenally about states of that type, or at least that the state wasn't then highlighted by attention, and the subject now can't remember it for that reason. This might seem to be fine for the Dispositional HOT theory, since in either case it would seem to follow that the subject wouldn't introspectively have judged the state to be conscious at the time, either because it was of a type entirely outside the realm of phenomenal access, or because the lack of attention meant that the subject couldn't then have formed an introspective phenomenal concept for it.

But now consider this complication. Suppose the earlier state was otherwise of a type that can be reported phenomenally, but was *pre-attentive*, by which I mean that it consisted of a kind of minimal sensory registration which isn't itself attention, but which can be worked up into attention by a match with some stored template. The trouble is then that, despite the negative memory report ('I wasn't consciously Øing'), due to the lack of earlier attention, it is arguably nevertheless be true that the subject *would* have reported the state to be conscious at the time, had the question come up. For the very raising of the question could have led the subject (a) to come to attend and thence (b) to form a phenomenal concept to characterize the experience (Am I hearing a sound? Am I seeing a bird in that tree?).

If this is right, then the Dispositional HOT theory falls out of kilter with the empirical methodology for investigating consciousness. For such pre-attentive states *will* be dispositionally Higher-Order-judgeable—the subject would introspectively have judged them to be present earlier, had the question been raised. Yet the subject's phenomenal memory reports will be negative—the earlier lack of attention will lead the subject to deny hearing a sound earlier, say, or seeing a bird in the tree. So dispositional Higher-Order judgeability will be *present* in some cases which the empirical methodology catalogues as *not* conscious, thus indicating that dispositional Higher-Order judgeability is not sufficient for consciousness after all.

Perhaps the Dispositional HOT theory can be patched. One possibility would be to argue that pre-attentive states are *not* really dispositionally Higher-Order-judgeable, perhaps on the grounds that the conditions required for introspective phenomenal reports about them are too demanding, in needing the creation of a match with some template, and not just the raising of a question about the subject's current phenomenal state.

Alternatively, Dispositional HOT theorists could insist that the earlier pre-attentive states *were* conscious at the time, and put down the later negative reports to failures of memory. This would be roughly analogous to a move dismissed earlier, when I argued that Actualist HOT theorists cannot reasonably discount all later *positive* phenomenal memory reports about earlier *unintrospected* experiences; here Dispositional HOT theorists want to dismiss *negative* phenomenal memory reports about earlier *pre-attentional* experiences. But perhaps the present move isn't as unreasonable as the earlier one, in that omissions of memory are generally easier to understand than confabulations of memory, and lack of attention seems a more plausible basis for memory failure than mere lack of introspection.

I do not propose to pursue this knotty issue any further. Let me now assume, for the sake of the argument, that the Dispositional HOT theory can somehow be engineered into line with the empirical data. Even given this assumption, the debate about materialist theories of consciousness-as-such would by no means be over. For one thing, there may still be other material properties which fit the empirical requirements for consciousness-as-such equally well. Moreover, as I shall show in the next section, there is room for some radical doubts about the cogency of these empirical requirements themselves.

Our earlier discussion of determinate phenomenal properties showed how a number of different material properties can all fit the same empirical requirements. The empirical methodology can leave us with more than one material candidate for the referent of some phenomenal concept. In the case at hand, where we are dealing with the determinable, consciousness-as-such, we will have alternative candidates for a material referent as soon as there is some other property, apart from dispositional Higher-Order judgeability, which fits the empirical data equally well.

Our earlier discussion of determinate phenomenal properties already points to some such alternatives. Thus there are the alternatives arising from the choice between structure and substance.

It is natural to take dispositional Higher-Order judgeability as a structural property, a higher property that can be shared by beings with different physical constitutions. But then this higher property will be realized in humans by various lower-level material properties, to do with the physical and physiological nature of human brains. These lower-level properties will thus be present whenever humans affirm the presence of consciousness, and absent whenever they deny it. So our empirical methodology will be unable to decide between these lower-level properties and dispositional Higher-Order judgeability as the material nature of consciousness-as-such.

More generally, we can see that any material feature which in normal humans correlates exactly with dispositional Higher-Order judgeability will not be eliminated as a candidate for the nature of consciousness-as-such by the standard methodology. Note that such a feature need not even metaphysically *determine* dispositional Higher-Order judgeability, as do the relevant physical or physiological features of human brains. It will be enough if it is fully correlated with dispositional Higher-Order judgeability in normal humans, even if it could occur without dispositional Higher-Order judgeability in other creatures.

Our earlier discussion of the match between dispositional Higher-Order judgeability and the empirical data makes it clear that there must be some such correlated, but non-Higher-Order, properties. Recall the issue of pre-attention. Dispositional Higher-Order theorists need *either* to say that pre-attentive states are conscious (and negative memory reports to the contrary are mistaken), *or* that pre-attentive states are not conscious (and that they are not dispositionally Higher-Order judgeable either). Well, on the former option, *pre-attention* will itself correlate perfectly in humans with dispositional Higher-Order judgeability: for, when you have a pre-attentive state, then you will, if you put your mind to it, make an introspective phenomenal judgement about that state; conversely, when a state is not even pre-attentive, there is no question of making such a phenomenal judgement. Alternatively, on the other option, where pre-attentive states are held to be neither conscious nor Higher-Order judgeable, then *attention* will come out perfectly correlated with dispositional Higher-Order judgeability: for, as before, if you have an attentive state, then you will make an introspective phenomenal judgement about it, if you put your mind to it; and now, if you don't have an attentive state, then you will be held not to be sufficiently well primed for the making of such an introspective judgement.

Note now how neither attention nor pre-attention themselves determine any kind of Higher-Order activity: a being can have attentive or pre-attentive states which are directed on the *world*, without being able to think about *experiences*. This shows that, on any version of a Dispositional Higher-Order theory, there will be some non-Higher-Order property which in normal humans goes hand in hand with dispositional Higher-Order judgeability, and which the empirical data will therefore favour equally as a candidate for material analysis of consciousness-as-such.

Just as before, I take our inability to decide empirically between all these alternative material analyses of consciousness-as-such to be a symptom of vagueness in our phenomenal concepts, rather than of any epistemological failing. Along with our more determinate phenomenal concepts, our phenomenal concept of consciousness-as-such is a crude tool, lacking theoretical articulation; its task, I take it, is simply to categorize subjects into those which have the kind of cerebral states that our phenomenal concepts enable us to recognize first-personally and those which lack such states. It will do this effectively enough if it hooks on to any of the package of different properties which are present whenever humans attest to consciousness, and absent whenever they disclaim it. Beyond that, there is no reason to expect it to point more specifically, to the property of dispositional Higher-Order judgeability, say, or to the physical or physiological properties which realize such availability in humans, or to attention or pre-attention, or to anything else which goes hand in hand with all these properties in humans.

7.14 *Methodological Meltdown*

In fact, I think that the phenomenal concept of consciousness-as-such may be even more vague than this.

I have just argued that this concept is indecisive between a number of properties which coincide in humans. Let me now divide these into two groups: (a) properties which fix dispositional Higher-Order judgeability, such as this property itself, or its physical basis in humans; (b) properties, like attention or pre-attention, which may happen to coincide with dispositional Higher-Order judgeability in humans, but which do not themselves determine this property.

Now suppose, for the sake of the argument, that we take the phenomenal concept of consciousness-as-such to refer to one of these latter (b) properties, like attention. One obvious upshot would be that consciousness can then be present in beings who are not capable of HOT judgements, for lack of any phenomenal concepts. In many people's eyes, this is a major attraction of such less demanding (b) analyses of consciousness-as-such. It seems clear that human infants under two, and nearly all animals, including many higher mammals, lack any phenomenal concepts with which to make HOT judgements about their own experiences. Yet it strikes many people as absurd to hold that one-year-old babies and cats lack all consciousness. Equating consciousness with attention, say, which infants and higher mammals presumably do have, rather than with something that requires Higher-Order thinking about experiences, thus has the virtue of allowing babies and cats to be conscious.

My aim in rehearsing this familiar point is not to argue for the less demanding (b) analyses of consciousness over the more demanding (a) ones. As I have said, I don't think that there is a fact of the matter here: the phenomenal concept of consciousness-as-such is vague between these options. Rather, my concern is solely to point out that nothing so far definitely rules out the possibility that consciousness may be present in beings who do not themselves have phenomenal concepts. And the reason I want to stress this possibility is that it raises even further questions about the precision of the concept of consciousness, and indeed about the cogency of the 'standard methodology' that I have been assuming so far.

Focus now on the possibility of beings who have conscious experiences but lack any phenomenal concepts with which to think phenomenally about those experiences. We might say that 'it is like something' for these beings to have experiences, even though they are not 'aware *of*' these experiences, in the sense of forming phenomenal judgements about them. However, if there can be such phenomenally unrecognized experiences in babies and cats, why shouldn't there also be similarly phenomenally unrecognized experiences in adult human beings? That is, what rules out the possibility that, alongside the experiences that our phenomenal concepts make us aware of, we have other 'hidden' experiences, which are equally conscious, but to which we have no first-personal access, for lack of any corresponding phenomenal concepts? I realize that this may seem a very odd suggestion. But it is hard to see why we should dismiss it outright, once we allow that there may be conscious experiences without corresponding phenomenal concepts, as in babies and animals.[18]

If we were to allow that there can be conscious states which are 'hidden' in this way from normal human subjects, then this would have radical implications for the empirical methodology for studying consciousness-as-such. The structure of this methodology is as follows. We postulate some X as the material nature of consciousness. Then we look at the positive cases, where humans report they *are* conscious, to check that X is always present, and so isn't shown to be unnecessary for consciousness. And we look at the negative cases, where humans say they are *not* conscious, to check that X is always absent, and so isn't shown to be insufficient for consciousness. But if we allow that conscious states can be hidden from normal subjects, then the negative side of this methodology falls away. The fact that normal humans report themselves not to have first-person phenomenal knowledge of some state can no longer be taken to show that this state is not conscious. All we have left are the positive cases.

This would then have the surprising implication that the empirical

[18] It may be helpful to think here of hypnotically eliminated pains, or of multiple personality disorder, or of fugues. In all these cases, we have human subjects who deny awareness of states we would normally take to be conscious. Still, it is also true that these states are otherwise physically identical to states which in standard cases are first-person accessible to human subjects; moreover, in the non-standard cases we can often excavate a 'further subject' who does own to first-person knowledge of these states. I am also raising the even more extreme possibility of conscious states of a physical kind which are *never* first-person accessible to *any* human subjects.

methodology gives us no way of distinguishing among those many properties that are common to the positive cases where humans report that they *are* conscious. These properties will range from relatively specific properties, which are peculiar to such positive cases, such as dispositional Higher-Order judgeability, say, or attention, to much more general properties, which are common to the positive cases but not restricted to them, such as being implemented in connectionist structures, or being made of organic compounds, or even being material. Such general properties will be instanced in a far wider range of human states than are reported by humans as conscious. Yet the methodology now being countenanced would be impotent to rule them out as candidates for consciousness-as-such on these grounds. For the only requirement still left would be that a candidate property must be present when adult humans report positively that they *are* conscious. It would be no disqualification that it is also sometimes present when adult humans disclaim conscious awareness. So any property, however general—even being made of matter—would not be ruled out, as long as it is common to all cases where humans positively report conscious awareness.[19]

I am sure that some readers are feeling impatient with these radical methodological speculations, along with the underlying suggestion that there may be 'hidden' conscious experiences in adult humans. For one thing, you may want to say, don't conscious experiences have to be experiences *for* a subject, not somehow floating around

[19] Note that this weakening of the standard methodology need not affect research into determinate conscious properties, like seeing something red or hearing middle C, as much as it affects research into consciousness-as-such. For, even if we allow that conscious states can be hidden (including perhaps 'subliminal' versions of seeing colours and hearing notes), we might still have *some* negative phenomenal information about the absence of such specific modes of consciousness. This could come in the form of positive phenomenal judgements about related but incompatible phenomenal properties. Thus subjects might judge that they are seeing something green, which would rule out their seeing something red, or that they are hearing E, which would rule out their hearing middle C. The point is that such phenomenal reports will contain the information that some determinate phenomenal property is *not* present; by contrast, once we admit hidden states, negative reports on consciousness-as-such can no longer be read as saying that some state is definitely *not* conscious, only that it was not *judged* to be conscious.

unattached? Well, maybe so, but we need to be wary of building too much into this requirement. Given that we are already allowing that there may be conscious states in babies and cats, then we cannot be requiring that conscious states must be 'for a subject' in the sense that the subject can think *about* them. At most, the requirement is that conscious states must somehow be states of some persisting entity, that they must somehow contribute to some continuing whole. But it is not clear why any requirement of this kind should be violated by the possibility of hidden conscious experiences. Since 'hidden' here simply means opaque to Higher-Order judgement, it leaves it open that the experiences so hidden may still be 'for a subject' in some other sense. This then leaves it to empirical research to decide in exactly what such sense conscious experiences must be 'for a subject', and we are back where we started.

Again, some readers may feel that the notion of conscious states which are hidden from adult humans belies the very idea of consciousness. If there is anything definite about phenomenal consciousness, surely it is that we know when we are conscious. If a human cognitive state doesn't reveal itself to our first-personal phenomenal scrutiny, isn't this just to say that it is not a conscious state?

But I wonder how much this conviction rests on the dualist picture of consciousness as some extra, non-material spark which attaches to a special subclass of cognitive states. If you think of consciousness in this way, and think of phenomenal recognition as some system which internally scans the cognitive realm for this spark, then it may indeed seem odd that the spark of consciousness should be present, yet opaque to phenomenal recognition. But things look rather different if we think of phenomenal recognition not as requiring any extra, non-material spark, but simply as deriving from the existence of a phenomenal concept for the relevant state—that is, from an ability to place versions of that state inside the 'experience operator', and thereby to think about that state. If this is the basis for phenomenal recognition, then there is no obvious reason why there shouldn't be cerebral states which are similar to phenomenally recognizable states in all other important respects, but can't themselves be so recognized, for lack of any corresponding phenomenal concepts—in which event there would seem to be grounds for counting these states as conscious, even though 'hidden' from their introspecting subjects.

In any case, I have no interest in deciding this issue. This is another place where I think the phenomenal concept of consciousness-as-such goes fuzzy. I have already argued, in the last section, that there is no fact of the matter as to whether this concept refers to HOT-determining states of type (a), or more general states of type (b). I have now been arguing that if we regard type (b) referents as open, then there is a further choice, between taking the phenomenal concept of consciousness-as-such to cover states which are hidden from adult human observers, or taking it to be restricted to the sort of states which adult humans will first-personally recognize if they are present. I think there is no fact of the matter here either.

I said at the end of the last section that the phenomenal concept of consciousness-as-such is a crude tool, the purpose of which is to pick out people with the 'kind of cerebral states that our phenomenal concepts enable us to recognize first-personally'. But exactly what kind is this? A kind composed of states which are *inevitably* phenomenally recognizable when they occur in humans? Or a kind composed of states which are importantly similar to the those states which humans can recognize phenomenally, but which may also include 'hidden' states which cannot themselves be recognized phenomenally by humans? I don't see that there is anything in our phenomenal concept of consciousness-as-such to decide this issue.

I have now argued that phenomenal consciousness-as-such is vague in more than one dimension. Some readers may find this hard to credit. How can it be a vague matter whether some state is conscious or not? Surely it is either like something to have that state, or it is not. I shall say something about this gut reaction in section 7.16. But first let me briefly illustrate some of the points made in this section in connection with representational theories of consciousness.

7.15 *Representational Theories of Consciousness*

One popular approach to consciousness-as-such seeks to equate it with representation, or at least with certain kinds of representation (Harman 1990, Dretske 1995, Tye 1995, 2000). The thesis of such 'representational theories of consciousness' is that a creature is conscious just in case it is in a certain kind of representational state, some state which represents in a certain way.

From a phenomenal point of view, many conscious states certainly present themselves as manifestly representational. Thus to think of yourself phenomenally as *visually perceiving something* is to regard yourself as being in a state which represents the visible world as being a certain way. Similarly with other forms of sensory perception. When you think of yourself phenomenally as hearing, smelling, tasting, or touching something, you regard yourself as being in a state which represents the world as containing sounds or smells or tastes or textures located at various positions in space.

The same point applies to conscious *non-sensory thought*, as when I reflect, say, that the Roman Empire lasted more than seven centuries. Though I have not had occasion to consider non-sensory thought in this book so far, it seems clear that episodes of non-sensory thought can be conscious, and correspondingly that we can form phenomenal concepts of such episodes, as when we fill the gap in 'the experience: ---' with the conscious thought that the Roman Empire lasted for more than seven centuries.[20] And when we do think phenomenally about non-sensory thoughts in this way, then again we regard ourselves as being in states which represent things as being a certain way—for example, you might regard yourself as being in a state which represents the Roman Empire as having lasted for more than seven centuries.

Now, one question here is whether *all* conscious states strike us as similarly representational when we think about them

[20] Interestingly, where imaginative phenomenal references to *perceptual* states activate 'faint copies' of those states, imaginative phenomenal references to non-sensory thought episodes activate those thoughts themselves. When I imagine thinking that the Roman Empire lasted more than seven centuries, I think this very thought, not some faint copy of it.

phenomenally. However, from the point of view of this chapter, this is not the crucial issue. Our official concern here is not with how conscious states seem when thought about *phenomenally*. Rather, the question currently at issue is what *material* property, if any, is characteristic of all states that subjects report as conscious. It is an interesting enough question whether or not phenomenological reflection can deliver the verdict that the general run of conscious states all present themselves as representational from a phenomenal point of view. But, however this question pans out, it is a different question from whether those states all have some characteristic representational nature of a material sort—that is, some representational property understood in terms of a causal, teleosemantic, or similar materialistic theory of representation.

Of course, if all conscious states do appear representational from a phenomenal point of view, and if, in addition, all these states share some species of materially conceived representational property, then materialists will draw the conclusion that the phenomenal representational property is identical to the relevant species of material representation. This would simply be a special case of the kind of a posteriori mind-brain identity that has featured centrally in this book. We have a phenomenal concept of representation, which picks out consciousness-as-such phenomenally, and a material concept of representation, which empirical investigation shows to be coextensive with the phenomenal concept. So on this basis we conclude that the two concepts refer to the same property.

Still, as I said, my primary focus here is not with whether all conscious states appear representational from a phenomenal point of view, but rather with whether they share some representational property of the kind that might be articulated by a causal or teleosemantic account of representation. So I take the task facing a representational theory of consciousness to be as follows. Any such theory needs first to identify some material type of representation, of the kind discussed by causal or teleosemantic accounts of representation; it then needs to show, on the one hand, that all conscious states have this material property, and, on the other, that all states with this material property are conscious.

Let us take these two demands in turn. The first issue is whether all conscious states are appropriately representational. It is by no means obvious that this is so. Certainly there are some conscious states which are not obviously representational from a phenomenal point of view. For example, feelings of anger or sadness or drug-induced euphoria do not immediately present themselves phenomenally as representational. Still, as before, phenomenal appearance isn't the crucial issue. It could still be that these states turn out to be representational when analysed in terms of a materialist account of representation. Thus, for example, such a theory might imply that *anger* in fact represents that some injustice has been done, that *sadness* represents that things are generally going badly, and even that artificially induced *euphoria* represents that the world is a fine place.

I do not propose to spend time on this issue. For what it is worth, I see no special reason to suppose that all conscious states have a representational nature.[21] Why shouldn't some conscious states, like euphoria, say, simply feel like something, without represenating the world as being any particular way? True, in Chapter 4 I introduced my account of phenomenal concepts in a way that might have seemed to predetermine that all their referents are representational, in that I explained how versions of *perceptual* states could be embedded in an 'experience operator' to yield terms of the form 'the experience: ---', with the resulting terms then referring to *perceptual* experiences akin to the embedded states. And perceptual experiences are certainly very good candidates for being representational states. Still, as I explained in Chapter 4, I did not intend this to imply that all phenomenal concepts refer to perceptual states, or that they all refer to representational states. The focus on perceptual states was expositorily convenient, but nothing in my earlier discussion ruled out the possibility that versions of conscious states which are not representational, like drug-induced euphoria, say, can be embedded in the experience operator to form phenomenal concepts for those non-representational states.

[21] Moreover, even if there is some such common feature, standard empirical evidence will be impotent to decide whether it should be construed broadly or narrowly, or in terms of human physical realization or higher structure. Cf. sections 7.6–7.7 above.

Let me turn to the second demand on representational theories of consciousness: that of showing that all appropriately representational states are conscious. This too is problematic. The trouble here is that there is a great deal of activity in the brain which is representational, but which doesn't seem conscious. For example, in the early stages of human visual processing there are states that represent changes in the wavelength and intensity of light waves. But these states don't seem to be conscious. We don't take ourselves to be consciously aware of these properties of light waves, even though our brain is registering them.

The natural tactic for representationalists at this point is to raise the stakes, and specify that not all kinds of representation constitute consciousness. For example, it could be held that consciousness arises only when representation plays a special role in controlling actions, or when it interacts with other beliefs and desires in inferential ways, or some such.

Now, maybe some line of this kind can be made to work, though the difficulties should not be underestimated—recent psychological research shows that many processes which are hidden to phenomenal introspection harbour suprisingly sophisticated forms of representational cognition, and so may prove hard to exclude by any of the more obvious ways of raising the representational stakes. (Cf. Goodale and Milner 1992, Weiskrantz 1986.) However, rather than pursue these difficulties, let me simply point out that, from the point of view argued in the last section, there is no compelling reason why representationalists should *need* to raise the stakes. For they have the alternative of holding that the troublesome sub-phenomenal representational states are conscious, even though they are hidden from phenomenal judgement. This would then free them from any need to raise the requirements for being appropriately representational. They could simply equate consciousness with representation in general, while noting that there are some human states which are so conscious, but are hidden from human phenomenal judgement.

Perhaps, though, it is unsurprising that advocates of representational theories of consciousness should fail to defend them in this way. For, as I pointed out in the last section, once you allow that there may be conscious states which are 'hidden' to normal human observers, then the methodology for studying consciousness loses nearly all its bite, and ceases to be able to discriminate between all the many properties that are common to the cases which humans report positively as conscious. Maybe these cases are all representational. But equally, they will all be connectionist structures, or made of organic compounds, or even material. Once we stop reading negative reports as telling us definitely that some state is *not* conscious, we have no way of ruling out any of these properties as the material essence of consciousness-as-such.

So the option of defending a representational theory of consciousness by appealing to the possibility of phenomenally hidden conscious states is something of a two-edged sword. This strategy may allow you to account for the fact that even some sophisticated forms of cerebral representation are opaque to phenomenal scrutiny: you can maintain that, even though these states are hidden from the introspective subject, they are still conscious. But at the same time this strategy effectively undermines the enterprise of identifying the material essence of consciousness-as-such in the first place, since it means that you cannot rule out connectionist structure, say, or organic constitution, or even materiality, as candidate essences for consciousness. In short, once you allow hidden states, then the possibility of serious theorizing about consciousness-as-such collapses. If you are going to defend a representational theory of consciousness in this way, then you probably shouldn't be in the business of theorizing about consciousness in the first place.

7.16 *Vagueness and Consciousness-as-Such*

I have argued that the phenomenal concept of consciousness-as-such is vague in more than one dimension. For a start, it is indeterminate whether it refers to dispositional Higher-Order judgeability or to any of the other correlated properties which are similarly present whenever humans report themselves conscious and absent whenever they deny this (such as the physical basis for Higher-Order judgeability in humans or attention or pre-attention).

Moreover, once we allow in this way that there may be consciousness without Higher-Order thought, then it is hard to rule out hidden conscious states in humans, in which case the phenomenal concept of consciousness-as-such seems to become indeterminate between all the many properties which are present whenever humans report themselves conscious, even if not absent whenever humans deny this (such as connectionist structure or organic constitution or materiality).

Still, as I said above, these imputations of vagueness are likely to strike many readers as absurd. Can it really be a vague matter whether some creature is conscious-as-such? Surely, many will want to insist, it is either *like something* for the creature, or it is not. How can this be a vague matter?

Vagueness for consciousness-as-such seems even more counter-intuitive than vagueness for determinate conscious concepts. You may have been persuaded by my earlier arguments that the phenomenal concept of *seeing something red*, say, is vague, in that it is indefinite whether the state of a silicon doppelganger looking at a ripe tomato is sufficiently like the corresponding human state to qualify as *seeing something red*. This thesis in itself isn't so weird: it allows, after all, that it may still definitely be like *something* for the doppelganger, and claims only that it is indefinite which specific human phenomenal category that 'something' falls under. But now I am arguing that it is indefinite whether it is like anything for the doppelganger in the first place. And this seems much harder to understand.

As it happens, the arguments of the last two sections have urged that there is rather *more* vagueness in the determinable phenomenal concept of consciousness-as-such than in determinates like the concept of seeing something red. This is because the possibility of hidden conscious states radically multiplies the alternative candidates for the material referent of the determinable concept, but not necessarily for the determinates (cf. n. 19 above). But let us pass over this complication. The issue I want to address in this section arises as soon as we allow *any* vagueness in the phenomenal concept of consciousness-as-such. In particular, it will arise even if you are unpersuaded by my hypothesis of hidden conscious states, but concede my prior point that the concept of consciousness-as-such is in any case indefinite between the various alternative properties that are both present whenever humans report themselves conscious and absent whenever they deny this. These properties, as I explained, will include dispositional Higher-Order judgeability, and the physical set-up which realizes such judgeability in humans, and attention or pre-attention, and indeed anything else which goes hand in hand with these properties in humans.

As soon as you allow even this much indeterminacy in consciousness-as-such, then the counter-intuitive implications follow. To keep it simple, consider my neighbour's cat Moggy. If consciousness-as-such consists in availability for HOT judgements, then Moggy is not conscious. But if it consists in attention, then Moggy is conscious. I say that the phenomenal concept of consciousness-as-such is indefinite between (at least) these two different referents. So it follows that there is no fact of the matter of whether Moggy is conscious.

Can I really say this? Well, let me repeat a point I made in connection with determinate phenomenal concepts. My claim is not that it is vague how it is for Moggy. There is nothing indefinite about the being of the cat. Rather, my thesis is that our phenomenal concept, conscious-as-such, is not precise enough to decide whether Moggy falls under it or not. There is nothing in the semantic constitution of this term which is able to determine whether or not it includes cats.

Still, some will insist, mustn't it either be *like something* for Moggy, or not? But I am not convinced that the mere phrase 'like something' will bear the weight of this argument. There are many ways of being, from those of humans who make phenomenal judgements about their own states, through cats who can attend but not introspect, down to amoebas and plants with simple sensorimotor systems. Why suppose that the phrase 'like something' draws a sharp line across this spectrum?

Of course, you will suppose this if you think that this phrase points to some separate kind of property, ontologically distinct from all material properties. For on this dualist view there will indeed be a clear difference between beings who have this extra kind of property and those which don't. But once we reject dualism, this thought falls away. There are many different kinds of material system, and no reason to think that a crude concept like the phenomenal concept of consciousness-as-such can sort them neatly into two kinds.

As I have stressed throughout this book, it is very hard to free ourselves from the dualist view. The intuition that phenomenal properties are distinct from any material properties is well-nigh inescapable. In my view, this is why we find it so hard to accept that consciousness is a vague matter. We assume that its being 'like something' involves some extra, non-material spark, and so conclude that either it is like something or it is not—either the spark is present, or it isn't. But if there aren't any such non-material sparks anywhere, then this a bad reason for thinking that there is always a precise fact of the matter about consciousness.

7.17 *Conclusion*

Where does all this leave the prospects for the scientific investigation of consciousness? There is no doubt that such investigation can tell us much that is interesting. In particular, it can identify processes which are present in human beings whenever they subjectively report the presence of some phenomenal property and absent whenever they deny this. Findings of this kind are often extremely surprising. In particular, it turns out that many of these processes are far more specific than might initially have been supposed: much high-level cognitive processing that has an influence on subsequent behaviour, and which we might therefore have expected to manifest itself as conscious, turns out not to be phenomenally accessible (for example, see Weiskrantz 1986, Goodale and Milner 1992, Libet 1993). And there are also converse cases, of surprising *positive* reports on phenomenal properties which we might initially have expected to be subjectively unavailable (for example, see Dennett 1978*b*).

Still, interesting as this research is, I have argued in this chapter that it is fated to deliver less than it promises. If you are hoping to put your finger on some specific material property which is guaranteed to make its possessor feel *like this* (and here you think phenomenally about pain, say), or even to put your finger on some specific material property which is guaranteed to make its possessor feel *like anything* at all (and here you think phenomenally about consciousness-as-such), then the scientific study of consciousness is going to fail you. For, given any phenomenal concept, there are many different material properties that are present whenever humans apply that concept first-personally and absent whenever they deny it, and scientific research will therefore be unable to discriminate between them. Yet such discrimination is needed if we are to be able to tell, of creatures in general, as opposed to humans in particular, when they feel *like this*, or *like anything* at all. (Moreover, it is not even clear that candidates for the material nature of some phenomenal property need to be absent whenever humans subjectively report its absence, which weakens the ability of scientific investigation to discriminate between such candidates even further.)

Perhaps it would have been helpful if I had been emphasizing more fully that these pessimistic conclusions apply specifically to *phenomenal* concepts, as opposed to psychological ones. As I explained in Chapter 4, I take everyday words for experiences, like 'pain' or 'seeing something red' or indeed 'conscious', to express psychological concepts, which pick out experiences via descriptions of their causal roles, as well as phenomenal concepts, which identify experiences in terms of how they feel. I have argued that phenomenal concepts are vague, and that scientific research is correspondingly unable to identify precise referents for them. But that this is true of phenomenal concepts does not mean that it must also be true of psychological concepts.

So some psychological concepts may be precise, even where their phenomenal counterparts are vague. Perhaps our psychological concept of pain, say, or even our psychological concept of consciousness-as-such, refers precisely to some definite material property, where the phenomenal concepts of pain or consciousness-as-such do not.

On this topic I have said nothing at all, and this is scarcely the place to start. It raises many large issues, which could well provide the material for another book. For what it is worth, I suspect that there is much vagueness in our psychological concepts too. At the same time, I have no doubt that on some points they will be precise where our phenomenal concepts are vague. To take one example that has figured prominently in this chapter, it seems clear to me that if we ever came across a silicon doppelganger, we would quickly come to regard it as conscious, and treat it accordingly. (We might continue to wonder whether its red experiences were the same as ours, but we would surely soon cease to doubt that it was conscious-as-such.) However, I take it that this conclusion would involve our psychological thinking about consciousness, not our phenomenal thinking. Whatever the exact logic that drives the conclusion, it will derive from considerations relating to the causal role of the doppelganger's states, not from direct investigation of whether it is *like anything* for the creature.

So perhaps our psychological concepts can draw lines where our phenomenal concepts are indecisive. This does not affect the moral of this chapter. As I have urged throughout this book, phenomenal and psychological concepts are a priori distinct. So any precision in a psychological concept will not automatically transfer itself to its phenomenal counterpart. Since it is a posteriori whether a given phenomenal concept refers to the same thing as some psychological concept, definiteness in a psychological concept will not remove vagueness in a corresponding phenomenal concept, if it is already vague whether they co-refer. (This means that the ready acceptance of the silicon doppelganger as conscious will be challengeable by those who distinguish phenomenal from psychological thinking: 'Sure it seems conscious, but can we be sure that it *feels like anything*, given that it lacks the physical properties present when we know *we* feel like something?' I take the fact that such quibbles would be unlikely to affect our personal dealings with the doppelganger to indicate the relative importance of phenomenal and psychological thinking in practical life, as opposed to theoretical reflection.)

It may be that much of the current enthusiasm for 'consciousness studies' has been fomented by a failure to separate phenomenal and psychological issues. The subject seems *exciting* because it promises to identify the material nature of feelings—it promises to pinpoint those material properties that constitute feeling *like this*, or *like anything* at all. At the same time, the subject seems *fruitful*, because there is plenty of room for progress in finding out when specified causal roles are satisfied in different creatures, and by what mechanisms. So the failure to distinguish sharply between phenomenal and psychological issues makes the study of consciousness seem simultaneously exciting and fruitful. But you can't have it both ways. If you are really after the excitement of the phenomenal questions, then you won't get the answers you are looking for. And if you really want the fruitful answers that can indeed be delivered by straightforward psychological research, then you shouldn't deceive yourself into thinking that they are settling the phenomenal questions.

I don't want to be a killjoy. As I said, the scientific study of consciousness has delivered many interesting findings, and will no doubt continue to do so. But we need to see it for what it is. It will serve no good purpose to pretend that it can resolve phenomenal questions that are in fact unanswerable. There is nothing wrong with ambition. But there is no virtue in aiming for illusory goals.

Appendix: The History of the Completeness of Physics

> The flood of projects over the last two decades that attempt to fit mental causation or mental ontology into a 'naturalistic picture of the world' strike me as having more in common with political or religious ideology than with a philosophy that maintains perspective on the difference between what is known and what is speculated. Materialism is not established, or even deeply supported, by science. (Burge 1993: 117)

> No one could seriously, rationally suppose that the existence of antibiotics or electric lights or rockets to the moon disproves . . . mind–body dualism. But such achievements lend authority to 'science', and science . . . is linked in the public mind with atheistic materialism. (Clark 1996)

A.1 *Introduction*

Those unsympathetic to contemporary materialism sometimes like to suggest that its rise to prominence since the middle of the twentieth century has been carried on a tide of fashion. On this view, the rise of physicalism testifies to nothing except the increasing prestige of physical science in the modern *Weltanschauung*. We have become dazzled by the gleaming status of the physical sciences, so the thought goes, and so foolishly try to make our philosophy in its image.

I reject this suggestion. In Chapter 1 I explained how materialism follows from a serious argument with persuasive premisses. Moreover, as I also intimated in Chapter 1, a proper appreciation of this argument indicates an alternative explanation of why philosophical materialism has only recently become so widespread. A crucial premiss in the causal argument is the completeness of physics, and this premiss lacked convincing empirical support until well into the twentieth century. The reason why earlier philosophers were not materialists is not that they lacked some scientistic prejudice peculiar to the later twentieth century (after all, there were plenty of enthusiasts for science in previous centuries), but simply that they lacked the evidence which has now persuaded modern science of the completeness of physics.

In this Appendix I want to rehearse the history of scientific attitudes to the completeness of physics, and to show how changing views about this claim have interacted with attitudes to the mind–body problem. This will confirm my suggestion that modern materialism flows from the recent availability of the completeness of physics. But before I proceed, let me make one preliminary point. My claim that materialism derives from the completeness of physics might seem to be belied by the fact that few of the philosophers who developed modern materialism in the middle of the twentieth century, like the Australian central state materialists or David Lewis or Donald Davidson, made any explicit reference to this principle. If I am right that the completeness of physics was the crucial new factor, then we might have expected these philosophers to say so.

However, this is a relatively superficial worry. It is true that these founding fathers of modern materialism offered a number of variant arguments for materialism, and that not all of these arguments feature the completeness of physics as prominently as does the causal argument detailed in Chapter 1. Even so, it is not hard to see that nearly all these other arguments presuppose the completeness of physics in one way or another, and would not stand up without it. The original defenders of materialism in the middle of the twentieth century may not have been explicit about the importance of the completeness of physics, but it remains the case that their innovatory views would not have been possible without it.

Thus, for example, consider J. J. C. Smart's (1959) thought that we should identify mental states with brain states, for otherwise those mental states would be 'nomological danglers' which play no role in the explanation of behaviour. Similarly, reflect on David Lewis's (1966) and David Armstrong's (1968) argument that, since mental states are picked out by their a priori causal roles, including their roles as causes of behaviour, and since we know that physical states play these roles, mental states must be identical with those physical states. Or again, consider Donald Davidson's (1970) argument that, since the only laws governing behaviour are those connecting behaviour with physical antecedents, mental events can only be causes of behaviour if they are identical with those physical antecedents.

Now, these are all rather different arguments, and they give rise to rather different versions of materialism. But the point I want to make here is not sensitive to these differences. It is simply that none of these arguments would seem even slightly plausible without the completeness of physics. To see this, imagine that the completeness of physics were not true, and that some physical effects (the movement of matter in arms, perhaps, or the electrochemical changes which instigate those movements) were not determined by law by prior physical causes at all, but by *sui generis* non-physical mental causes, such as decisions, say, or exercises of will, or perhaps just pains. Then (1) contra Smart, mental states wouldn't be 'nomological danglers', but directly efficacious in the production of behaviour; (2) contra Armstrong and Lewis, it wouldn't necessarily be physical states which played the causal roles by which we pick out mental states, but quite possibly the *sui generis* mental states themselves; and (3) contra Davidson, it wouldn't be true that the only laws governing behaviour are those connecting behaviour with physical antecedents, since there would also be laws connecting behaviour with mental antecedents.[1]

A.2 *Descartes and Leibniz*

Let us now focus on the history of the completeness of physics. It may seem at first sight that the completeness of physics will follow from any version of physical theory which is formulated in terms of conservation laws. If the laws of mechanics tell us that important physical quantities are conserved whatever happens, then doesn't it follow that the later physical states of a system will always be fully determined by their earlier physical states?

Not necessarily. It depends on what conservation laws you are committed to. Consider Descartes's mechanics. This incorporated the conservation of what Descartes called 'quantity of motion', by which he meant mass times speed. That is, Descartes held that the total mass times speed of any collection of bodies is guaranteed to remain constant, whatever happens to them. However, this alone does not guarantee that physics is complete. In particular, it does not rule out the possibility of physical effects that are due to irreducibly mental causes.

This is because Descartes's *quantity of motion* is a non-directional (scalar)

[1] In other writings from the middle of the century, the relevance of the completeness of physics does not need to be exacavated, since it lies on the surface. Thus see Feigl 1958, Oppenheim and Putnam 1958.

quantity, defined in terms of speed, as opposed to the directional (vectorial) Newtonian notion of linear *momentum*, defined in terms of velocity. Because of this, the *direction* of a body's motion can be altered without altering its quantity of motion. As Roger Woolhouse explains the point, in an excellent discussion of the relevance of seventeenth-century mechanics to the mind–brain issue (1985), a car rounding a corner at constant speed conserves its 'quantity of motion', but not its momentum.

This creates room for non-physical causes, and in particular *sui generis* mental causes, to alter the *direction* of a body's motion without violating Descartes's conservation principle. That principle does mean that if one physical body starts going faster, this must be due to another physical body going slower. But his principle doesn't require that if a physical body changes direction, this need result from any other physical body changing direction. Even if the change of direction results from an irreducibly mental cause, the quantity of motion of the moving body remains constant.

According to Leibniz, Descartes exploited this loophole to explain how the mind could affect the brain. As Leibniz tells the story, Descartes believed that the mind nudges moving particles of matter in the pineal gland, causing them to swerve without losing speed, like the car going round the corner, and then used this to explain how the mind could affect the brain without violating the conservation of 'quantity of motion' (Leibniz, 1898 [1696]: 327).

Now, there is little evidence that Descartes actually saw things this way, or indeed that he was particularly worried about how the laws of physics can be squared with mind–brain interaction. Still, whatever the truth of Leibniz's account of Cartesian theory, his next point deserves our attention. For Leibniz proceeds from his analysis of Descartes to the first-order assertion that the *correct* conservation laws, unlike Descartes's conservation of quantity of motion, *cannot* in fact be squared with mind–body interaction.

Leibniz's conservation laws were in fact a great improvement on Descartes's. In place of Descartes's conservation of 'quantity of motion', Leibniz upheld both the conservation of linear *momentum* and the conservation of kinetic *energy*. These two laws led him to the correct analysis of impacts between moving bodies, a topic on which Descartes had gone badly astray.[2] And, in connection with the mind–body issue, they

[2] Leibniz took it that all basic material particles are perfectly elastic, and that no kinetic energy is lost when they collide. He explained the apparent loss of kinetic energy when inelastic *macroscopic* bodies collide by positing increased motion in the microscopic parts of those bodies. (Thus he explains, in the fifth paper of the Leibniz-Clarke Correspondence, H. Alexander (ed.), 1956: 'The author objects,

persuaded him that there is no room whatsoever for mental activity to influence physical effects.[3]

In effect, the conservation of linear momentum and of kinetic energy together squeeze the mind out of the class of events that cause changes in motion. Leibniz's two conservation laws, plus the standard seventeenth-century assumption of no physical action at a distance, are themselves sufficient to fix the evolution of all physical processes. The conservation of momentum requires the preservation of the same total 'quantity of motion' in *any given direction*, thus precluding any possibility of mental nudges altering the direction of moving physical particles. Moreover, the conservation of energy, when added to the conservation of momentum, fully fixes the speed and direction of impacting physical particles after they collide. So there is no room for anything else, and in particular for anything mental, to make any difference to the motions of physical particles, if Leibniz's two conservation laws are to be respected.

We can simplify the essential point at issue here by noting that Leibniz's conservation laws, unlike Descartes's, ensure physical determinism, in the sense of implying that the physical states of any system of bodies at one time fix their state at any later time. Physical determinism in this sense is certainly sufficient for the completeness of physics, even if the possibility of quantum-mechanical indeterminism means that it is not necessary (cf. ch. 1 n. 2). So Leibniz's dynamics, unlike Descartes's, makes it impossible for anything except the physical to make a difference to anything physical.

Leibniz was fully aware of the implications of his dynamical theories for mind–body interaction (cf. Woolhouse 1985). However, he did not infer mind–brain identity from his commitment to the completeness of physics. Instead, he adopted the doctrine of pre-established harmony, according to which the mental and physical realms are each causally closed, but pre-arranged by the divine will to march in step in such a way as to display the standard mind–brain correlations. In terms of the causal argument laid out in Chapter 1, Leibniz is denying the first premiss, about the causal influence of mind on matter. He avoids identifying mental causes with

that two soft or un-elastic bodies meeting together, lose some of their force. I answer, no. 'Tis true, their wholes lose it with respect to their total motion; but their parts receive it, being shaken (internally) by the force of their concourse.')

[3] I am here using 'physical' in the sense it would have been understood by seventeenth-century mechanical philosophers, as referring to primary properties like mass and motion, and to anything ontologically determined thereby. In the next section I shall consider the Newtonian system, which allows an open-ended range of primary properties; in that context I shall use 'physical' to refer more generally to those properties which are 'inanimate' in the sense of Chapter 1.

physical causes, in the face of the completeness of physics, by denying that mental causes ever have physical effects.

A.3 *Newtonian Physics*

Some readers might now be wondering why this wasn't the end of the story. Given that Leibniz established, against Descartes, that both momentum and energy are conserved in systems of moving particles, why wasn't the history of the mind-brain argument already over? Of course, we mightn't nowadays want to follow Leibniz in opting for pre-established harmony, as opposed to simply embracing mind-brain identity. But this would simply be because we favour a different response to the causal argument laid out in Chapter 1, not because we have any substantial premisses that Leibniz lacked. In particular, the crucial second premiss of the causal argument, the completeness of physics, would seem already to have been available to Leibniz. So doesn't this mean that everything needed to appreciate the causal argument was already to hand in the second half of the seventeenth century, long before the rise of twentieth-century materialism?

Well, it was—but only on the assumption that Leibniz gives us the correct dynamics. However, Leibniz's physical theories were quickly eclipsed by those of Newton, and this then reopened the whole issue of the completeness of physics.

The central point here is that Newton allowed forces other than impact. Leibniz, along with Descartes and all other pre-Newtonian proponents of the 'mechanical philosophy', took it as given that all physical action is by contact. They assumed that the only possible cause of a change in a physical body's motion is the impact of another physical body. (Or more precisely, as we are telling the story, Descartes supposed that the only possible *non-mental* cause of physical change is impact, and Leibniz then argued that *mental* causes other than impact are not possible either, if the conservation of momentum and energy are to be respected.)

Newtonian mechanics changed the whole picture. This is because Newton did not take impact as his basic model of dynamic action. Rather, his basic notion was that of an *impressed force*. Rather than thinking of 'force' as something inside a body which might be transferred to other bodies in impact, as did all his contemporaries (and indeed most of his successors for at least a century[4]), Newton thought of forces as disembodied entities, acting on the affected body from outside. An impressed force

[4] Cf. Papineau 1977.

'consists in the action only, and remains no longer in the body when the action is over'. Moreover, 'impressed forces are of different origins, as from percussion, from pressure, from centripetal force' (Newton 1966 [1686]: 2, definition IV). Gravity was the paradigm. True, the force of gravity always arose from the presence of massive bodies, but it pervaded space, acting on anything that might be there, so to speak, with a strength as specified by the inverse square law.

Once disembodied gravity was allowed as a force distinct from the action of impact, then there was no principled barrier to other similarly disembodied special forces, such as chemical forces or magnetic forces or forces of cohesion (cf. Newton 1952 [1704]: queries 29–31)—or indeed vital and mental forces.

Nothing in classical Newtonian thinking rules out special mental forces. While Newton has a general law about the effects of his forces (they cause proportional changes in the velocities of the bodies they act on), there is no corresponding general principle about the causes of such forces. True, gravity in particular is governed by the inverse square law, which fixes gravitational forces as a function of the location of bodies with mass. But there is no overarching principle dictating how forces in general arise. This opens up the possibility that there may be *sui generis* mental forces, which would mean that Newtonian physics, unlike Leibnizian physics, is not physically complete. Some physical processes could have non-physical mental forces among their causal antecedents. (Some readers may be feeling uneasy about the way in which the completeness of physics has now turned into an issue about what 'forces' exist. I shall address this issue at the end of this section.)

The switch from a pure impact-based mechanical philosophy to the more liberal world of Newtonian forces undermined Leibniz's argument for the completeness of physics. Leibniz could hold that the principles governing the physical world leave no room for mental acts to make a difference because he had a simple mechanical picture of the physical world. Bodies preserve their motion in any given direction until they collide, and then they obey the laws of impact. The Newtonian world of disembodied forces is far less pristine, and gives no immediate reason to view physics as complete.

You might think that the conservation laws of Newtonian physics would themselves place constraints on the generation of forces, in such a way as to restore the completeness of physics. But this would be a somewhat anachronistic thought. Conservation laws did not play a central role in Newtonian thinking—at least not in that of Newton himself and his immediate followers. True, Newton's mechanics does imply the conservation of *momentum*. This falls straight out of his Third Law, which requires that 'action and reaction' are always equal. But it is a striking feature of Newtonian dynamics that there is no corresponding law for energy.[5]

Of course, as we shall see in the next section, the principle of the conservation of kinetic *and* potential energy in all physical processes did *eventually* become part of the Newtonian tradition, and this does impose a general restriction on possible forces, a restriction expressed by the requirement that all forces should be 'conservative'. But this came much later, in the middle of the nineteenth century, and so had no influence on the range of possible forces admitted by seventeenth- or eighteenth-century Newtonians. (Moreover, it is a nice question, to which we shall return at length below, how far the principle of the conservation of kinetic plus potential energy, with its attendant requirement that all forces be conservative, does indeed constitute evidence against *sui generis* mental forces.)

In any case, whatever the significance of later Newtonian derivations of the conservation of energy, early Newtonians themselves certainly saw no barrier to the postulation of *sui generis* mental forces. In a moment I shall give some examples. But first it will be helpful to distinguish in the abstract two ways in which such a Newtonian violation of the completeness of physics could occur.

First, and most obviously, it could follow from the postulation of

[5] One barrier to the formulation of an energy conservation principle by early Newtonians was their lack of a notion of potential energy, the energy 'stored up' after a spring has been extended or compressed, or as two gravitating bodies move apart. Given this, there was no obvious sense in which they could view two gravitating bodies, for example, as conserving energy while they moved apart: after all, the sum of their kinetic energies would not be constant, but unequivocally decreasing. And even in the case of impact, where the notion of potential energy is not immediately needed, early Newtonians displayed no commitment to the conservation of (kinetic) energy. Most obviously, Newton and his followers were perfectly happy, unlike Leibniz, to allow unreduced inelastic collisions, in which both bodies lose kinetic energy without transmitting it to their internal parts. It is also worth remarking that there is nothing in Newton's Laws of Motion to rule out even 'superelastic' impacts, in which total kinetic energy increases. If two bodies with equal masses and equal but opposite speeds both rebounded after collison with double their speeds, for example, Newton's three Laws of Motion and the conservation of momentum would be respected. True, any such phenomenon would provide an obvious recipe for perpetual motion, but the point remains that Newton's Laws themselves do not rule it out. (It is also worth noting that perpetual motion was no by means universally rejected by seventeenth- and eighteenth-century physicists. Cf. Elkana 1974: 28–30.)

indeterministic mental forces. If the determinations of the self (or of the 'soul', as they would have said in the seventeenth and eighteenth centuries) could influence the movements of matter in spontaneous ways, then the world of physical causes and effects would obviously not be causally closed, since these spontaneous mental causes would make a difference to the unfolding of certain physical processes.

But, second, it is not even necessary for the violation of completeness that such *sui generis* special forces operate indeterministically. Suppose that the operation of mental forces were governed by fully *deterministic* force laws (suppose, for example, that mental forces obeyed some inverse square law involving the presence of certain particles in the brain). Then mental forces would be part of Newtonian dynamics in just the same sense as gravitational or electrical forces: we could imagine a system of particles evolving deterministically under the influence of all these forces, including mental forces, with the forces exerted at any place and time being deterministically fixed by the relevant force laws. Even so, this deterministic model would still constitute a violation of the completeness of physics, for the physical positions of the particles would depend *inter alia* on prior mental causes, and not exclusively on prior physical causes.

Did I not say at the end of the last section that determinism is sufficient for the completeness of physics (even if not necessary, because of quantum mechanics)? No. What I said was that *physical* determinism (the doctrine that prior *physical* conditions alone are enough to determine later physical conditions) is sufficient for the completeness of physics. However, we can accept determinism as such without accepting physical determinism, and so without accepting the completeness of physics. In particular, we can have a deterministic model in which *sui generis* mental forces play an essential role, and in which the physical sub-part is therefore not causally closed.

You might feel (indeed, might have been feeling for some time) that a realm of deterministic mental forces would scarcely be worth distinguishing from the general run of physical forces, given that they would lack the spontaneity and creativity that is normally held to distinguish the mental from the physical. And you might think that it is therefore somewhat odd to view them as violating the completeness of physics. I happily concede that there is something to this thought. But I would still like to stick to my terminology, as stipulated in Chapter 1, which defined the 'physical' as whatever can be identified without using specifically animate terminology—which then makes even deterministically governed *sui generis* mental forces come out 'non-physical', since they can't be so identified. This is the terminology which best fits with our original interest in the causal argument for physicalism. We don't want deterministic mental forces to be counted as consistent with the 'completeness of physics', precisely because *this* kind of 'completeness of physics' wouldn't be any good for the causal argument: if mental forces are *part* of what makes 'physics' complete, then we won't be able to argue from this that mental forces must be identical with some *other* (inanimate) causes of their effects.

So far I have merely presented the possibility of special Newtonian forces as an abstract possibility. However, the postulation of such forces was a commonplace among eighteenth-century thinkers, particularly among those working in anatomy and physiology. Many of the theoretical debates in these areas were concerned with the existence of vital and mental forces, and with the relation between them. Among those who debated these issues, we can find both the indeterministic and deterministic models of mental forces.[6]

Thus consider the debate among eighteenth-century physiologists about the relative roles of the forces of *sensibility* and *irritability*. This terminology was introduced by the leading German physiologist Albrecht von Haller, professor of anatomy at Göttingen from 1736. Haller thought of 'sensibility' as a distinctively mental force. 'Irritability' was a non-mental but still peculiarly biological power. ('What should hinder us from granting irritability to be a property of the animal *gluten*, the same as we acknowledge gravity and attraction to be properties of matter in general': Haller 1936 [1751]: 211) Haller took the force of sensibility to be under the control of the soul and to operate solely through the nerves. Irritability, by contrast, he took to be located solely in the muscle fibres.

In distinguishing the mentally directed force of sensibility from the more automatic force of irritability, Haller can here be seen as conforming to my model of *indeterministic* mental forces. Whereas the force of irritability is determined by prior stimuli and is independent of mental agency, the force of sensibility responds to the spontaneous commands of the soul.

Haller's model was opposed by Robert Whytt (1714–66) in Edinburgh. In effect, Whytt can be seen as merging Haller's distinct mental and vital forces, irritability and sensibility. On the one hand, Whytt gave greater power to the soul: he took it that a soul or 'sentient principle' is distributed throughout the body, not just in the nerves, and is responsible for all bodily activities, from the flow of blood and motion of muscles, to imagination and reasoning in the brain. But at the same time as giving greater power to this sentient principle, he also rendered its operations *deterministic*. He explicitly likened the sentient principle to the Newtonian force of gravity, and viewed it as a necessary principle which acts according to strict laws.

[6] Here I am closely following Steigerwald 1998: ch. 2.

Whytt can thus be seen as exemplifying my model of deterministic mental forces: the sentient principle is simply another deterministic Newtonian force, just like gravity, in that its operations are fixed by a definite force law (Whytt 1755).

At this point let me say something about the terminology of 'forces' that I have been using in discussing Newtonian physics. It may be natural to present Newtonian physics in terms of reified forces in this way, but it is not mandatory. The alternative is to view the circumstances which supposedly generate these putative forces as themselves the direct causes of any resulting accelerations, and to regard the talk of 'forces' as simply a useful calculating device.

In an earlier paper about the history of the completeness of physics (Papineau 2000), I claimed that this choice made no difference to the issues, on the grounds that those who dispense with 'forces' can simply replace the question of whether there are 'mental forces' with the question of whether mental initial conditions ever make a difference to accelerations. (Cf. McLaughlin 1992: 64–5.) But now I think that the situation is more complicated, and that the reification of forces arguably makes it harder to uphold the completeness of physics.

The complication arises in connection with deterministic *mental* forces which are generated by special *physical* circumstances—for example, by circumstances found specifically within the brains of sentient beings. If we accept these as reified mental forces, then they would seem to violate the completeness of physics, since it seems that they will be needed as *sui generis mental* factors in any sufficient story about the causes of the accelerations they generate. On the other hand, if we refuse to reify forces, then a full story about the causes of those accelerations need mention only the prior *physical* circumstances which supposedly generate these 'mental forces', and the completeness of physics would thus seem to be respected. In such a case, then, the reification of forces seems to lead to a violation of the completeness of physics, where a non-reification does not.

The point is that while the non-reifiers may need special *laws* about the accelerations that are generated by the special physical circumstances found inside sentient bodies, laws that do not follow from other laws about accelerations, the antecedents of these special laws will still be physical, and so such antecedent causes will not violate the completeness of physics. By contrast, since those who reify forces do introduce *sui generis* mental forces to serve as causes of the relevant accelerations, their analysis of such accelerations will run counter to the completeness of physics. In short, special accelerations inside sentient brains would seem to violate the completeness of physics *if* we reify forces, but not otherwise.

Fortunately, we can bypass this issue about the reification of forces here. This is because I shall be arguing that there are in any case no special accelerations inside brains, and so no reason, even for those who reify forces, to introduce special mental forces. So I shall be able to uphold the completeness of physics even on the assumption that forces should be reified. Such reification may make it harder to defend the completeness of physics. But, if there are in fact no special accelerations inside brains to motivate mental forces, then no such falsification will result.

Given this, I shall continue to talk in terms of 'forces' in what follows. Since this only makes my argumentative task harder, this will give me no unfair dialectical advantage.[7] It simply sets me the greater challenge of showing that accelerations inside brains follow, not just from physical antecedents, but also in a way which is predictable from laws which also operate outside brains. (Or, as we would put it in terms of forces, that there are no forces inside brains which are not composed of physical forces which also operate outside brains)[8].

A.4 *The Conservation of Energy*

In this section I want to consider how the principle of the conservation of energy eventually emerged within the tradition of Newtonian mechanics, and how this bears on the completeness of physics.

[7] Moreover, it is probably the right way to talk anyway. While early Newtonian physics can avoid reifying forces, modern field theories cannot.

[8] Note that those who don't reify forces will be able to run a completeness-based causal argument for mind–brain identity *even if* cerebral accelerations are governed by special laws independent of other laws about accelerations. In this connection, it is useful to distinguish between 'weak' and 'strong' reduction. A weak reduction requires only that mental causes be identified with physical causes. A strong reduction requires also that the laws by which such causes operate follow by composition from non-special laws. Those who admit special cerebral accelerations but don't reify forces will only be able to achieve a weak reduction. If we do reify forces, on the other hand, then there is no room for a reduction which is weaker than a strong reduction: for once we introduce forces, there will be mental causes (forces) which are distinct from physical causes as soon as there are special mental laws which do not follow by composition from non-special laws. As I said, I myself have no need to make space for weak reduction by avoiding forces, since I shall argue that there are no special mental laws, and thus that stong reduction can be upheld anyway. (I think I remember May Brodbeck making this distinction between weak and strong reduction in the 1960s; but I haven't been able to track down a reference.)

A.4.1 *Rational Mechanics*

Through the eighteenth and early nineteenth centuries a number of mathematician-physicists, among whom the most important were Jean d'Alembert (1717–83), Joseph Louis Lagrange (1736–1813), the Marquis de Laplace (1749–1827), and William Hamilton (1805–65), developed a series of mathematical frameworks designed to simplify the analysis of the motion of interacting particles. These frameworks allowed physicists to abstract away from detailed forces of constraint, such as the forces holding rigid bodies together, or the forces constraining particles to move on surfaces, and to concentrate on the effects produced by other forces. (See Elkana 1974: ch. 2 for the history, and Goldstein 1964 for the mathematics.)

These mathematical developments also implied that, under certain conditions, the sum of kinetic energy and potential energy remains constant. Roughly, when all forces involved are independent of the velocities of the interacting particles and of the time (let us call forces of these kinds 'conservative'), then the sum of actual kinetic energy (measured by $\frac{1}{2}\Sigma mv^2$) plus the potential to generate more such energy (often called the 'tensions' of the system) is conserved: when the particles slow down, this builds up 'tensions', and if those 'tensions' are expended, the particles will speed up again.

We now think of this as the most basic of all natural laws. But this attitude was no part of the original tradition in rational mechanics. There were two reasons for this. First, the Newtonian scientists in this tradition were not looking for conserved quantities anyway. As I explained earlier, conservation principles played little role in classical Newtonian thinking. True, Leibniz himself had urged the conservation of kinetic energy (under the guise of 'vis viva'), but by the eighteenth century Leibniz's influence had been largely eclipsed by Newton's. Second, the conservation of potential and kinetic energy in any case holds only under the assumption that all forces are conservative. We nowadays take this requirement to be satisfied for all fundamental forces. But again, this was no part of eighteenth-century thinking. Some familiar forces happen to be conservative, but plenty of other forces are not. Gravitation, for example, is conservative, since it depends only on the positions of the particles, and not on their velocities, or the elapsed time. But, by contrast, frictional forces are not conservative, since they depend on the velocity of the decelerated body relative to the medium. And, correspondingly, frictional forces do not in any sense seem to conserve energy: when they decelerate a body, no 'tension' is apparently built up waiting to accelerate the body again.

For both these reasons, the tradition in rational mechanics did not initially view the conservation of kinetic and potential energy in certain systems as of any great significance. On the contrary, it was simply a handy mathematical consequence which falls out of the equations when the operative forces all happen to fall within a subset of possible forces. (Cf. Elkana 1974: ch 2.)

A.4.2 *Equivalence of Heat and Mechanical Energy*

In the first half of the nineteenth century a number of scientists, most prominently James Joule (1819–89), established the equivalence of heat and mechanical energy, in the sense of showing that a specific amount of heat will always be produced by the expenditure of a given amount of mechanical energy (as when a gas is compressed, say), and vice versa (as when a hot gas drives a piston).

These experiments suggested directly that some single quantity is preserved through a number of different natural interactions. They also had a less direct bearing on the eventual formulation of the conservation of energy. They indicated that apparently non-conservative forces like friction and other dissipative forces need not be non-conservative after all, since the kinetic energy apparently lost when they act will in fact be preserved by the heat energy gained by the resisting medium.[9]

The stage was now set for the formulation of a universal principle of the conservation of energy. We can distinguish three elements which together contributed to the formulation of this principle. First, the tradition of rational mechanics provided the mathematical scaffolding. Second, the experiments of Joule and others suggested that different natural processes all involve a single underlying quantity which can manifest itself in different forms. Third, these experiments also suggested that apparently non-conservative forces like friction were merely macroscopic manifestations of more fundamental conservative forces.

Of course, it is only with the wisdom of hindsight that we can see these different strands as waiting to be pulled together. At the time they were hidden in abstract realms of disparate branches of science. It took the genius of the young Hermann von Helmholtz (1821–94) to see the

[9] One model for this preservation was the kinetic theory of heat, which took the macroscopic kinetic energy apparently lost to be converted into internal kinetic energy at the microscopic level (cf. Leibniz's explanation for the apparent loss of kinetic energy in inelastic impact mentioned in footnote 2 above). But the abstract point at issue did not demand acceptance of the kinetic theory, since the lost kinetic energy could alternatively be viewed as being stored in the 'tensions' of whatever force might be associated with heat.

connections. In 1847, at the age of 26, he published his monograph *Über die Erhaltung der Kraft* ('On the Conservation of Force'). The first three sections of this treatise are devoted to the tradition of rational mechanics, and in particular to explaining how the total mechanical energy (kinetic plus potential energy) in a system of interacting particles is constant in those cases where all forces are familiar 'central forces' independent of time and velocity. The fourth section describes the equivalence between mechanical 'force' and heat, referring to Joule's results, while the last two sections extend the discussion to electric and magnetic 'forces', again showing that there are fixed equivalences between these 'forces', heat, and mechanical energy.[10]

A.4.3 *Physiology*

At the end of his treatise Helmholtz touches on the conservation of energy in living systems. Helmholtz was in fact a medical doctor by training, and had been a student in the Berlin physiological laboratory of Johannes Müller in the early 1840s, along with Emil Du Bois-Reymond (1818–96) and Ernst Brücke (1819–92). Together these students were committed to a reductionist programme in physiology, aiming to show that phenomena like respiration, animal heat, and locomotion could all be understood to be governed by the same laws as operate in the inorganic realm.

This physiological context undoubtedly played a fundamental role in Helmholtz's articulation of a universal principle of the conservation of energy. Because of his physiological concerns, Helmholtz was interested in a principle that would cover *all* natural phenomena, including those in living systems, and not just such manifestly physical phenomena as mechanical motion, heat, and electromagnetism. Thus he took the crucial step of asserting that *all* forces conserve the sum of kinetic and potential energy; superficially non-conservative forces like friction are simply macroscopic manifestations of more fundamental forces which preserve

[10] Helmholtz used the word 'Kraft'. This is now standardly translated as 'force' rather than 'energy', but these two concepts were not clearly distinguished at the time, neither in English nor German. The general expectation at the time was that any conservation law would involve 'force' ('Kraft', 'vis'), where this was thought of as a directed quantity ('force of motion'= momentum), rather than as a scalar like energy. (Here again we see the dominance of the Newtonian tradition, in which the only conserved quantity was the vectorial momentum.) One of Helmholtz's most important contributions was to make it clear that even within the Newtonian tradition of rational mechanics it is the scalar energy that is conserved, rather than any vectorial 'force'. Even so, the confusions persisted for some time, as shown, for example, by Faraday's 1857 paper 'On the Conservation of Force'. (Cf. Elkana 1974: 130–8.)

energy at the micro-level. This then enabled Helmholtz to view the equivalences established by experimentalists like Joule, not just as striking local regularities, but as necessary consequences of a fundamental principle of mechanics. All natural processes must respect the conservation of energy, including processes in living systems.

It seems likely that it was Helmholtz's specific combination of physiological interests and sophisticated physical understanding that precipitated his crucial synthesis of the different strands of research feeding into the conservation of energy. His desire to bring living systems under a unified science allowed him to see that if we assume that all fundamental forces are conservative, then this guarantees that a certain quantity, the total energy, will be preserved in all natural processes whatsoever, including the organic processes that formed the focus of his interest.[11]

A.4.4 *Vital Forces*

Helmholtz was part of a tradition in experimental physiology which set itself in opposition to the previous generation of German *Naturphilosophen*. During the eighteenth century the Newtonian categories of 'irritability' and 'sensibility' had gone through various transformations, and by the end of the century were widely referred to under the heading of *Lebenskraft*, or 'vital force', though there was continued disagreement on the precise nature of such forces. Meanwhile, within the tradition of German idealism, the notion of vital force had broken loose from its original Newtonian moorings, and became part of a florid metaphysics imbued with romanticism and idealism.

According to the *Naturphilosophen*, organic matter was infused with a special power which organized and directed it. Following Blumenbach and Kant, Schelling took up the term *Bildungstrieb* ('formative drive'), because of the excessively mechanical connotations he discerned in the traditional term *Lebenskraft*. Schelling and the other *Naturphilosophen* viewed this formative drive as having a quasi-mental aspect, which enabled it to mediate between the 'archetypical ideas' or 'essences' of different species and the development of individual organisms towards that ideal form. (See Coleman 1971: ch. 3; Steigerwald 1998.)

The experimental tradition which included Helmholtz can be seen as a

[11] In Papineau (2000) I suggested that their lack of physiological interests meant that the experimenters investigating mechanical equivalences, like Joule, had no interest in articulating a universal principle governing all natural interactions. However, this is wrong about Joule, who did defend such a universal principle. (Cf. Smith 1998: ch. 4.)

reaction to these extravagant doctrines. However, it is striking that many of those associated with this tradition, though not Helmholtz himself, continued to admit the possible existence of vital forces, both before and after the emergence of the conservation of energy. This is less puzzling than it might at first seem. These physiological thinkers did not think of vital forces as the mystical intermediaries of the *Naturphilosophen*, imbued with all the powers of creative mentality. Rather they were reverting to the tradition of eighteenth-century physiology. They viewed vital forces simply as special Newtonian forces, additional to gravitational forces, chemical forces and so on, which happen to arise specifically in organic contexts. Justus von Leibig (1803–73), the leading physiological chemist of the time, and Müller, Helmholtz's own mentor, are clear examples of experimental physiologists who were prepared to countenance vital forces in this sense. (Cf. Coleman 1971: ch. 6; Elkana 1974: ch. 4.)

A.4.5 *Does the Conservation of Energy Rule out Vital (and Mental) Forces?*

The interesting question, from our point of view is how far this continuing commitment to vital forces is consistent with the doctrine of the conservation of energy. There is certainly some tension between the two doctrines. It is noteworthy that Helmholtz himself, and his young colleagues from Müller's laboratory, were committed to the view that no forces operated inside living bodies that are not also found in simpler physical and chemical contexts (Coleman 1971: 150–4). Even so, there is no outright inconsistency between the conservation of energy and vital forces, and many late nineteenth-century figures were quite explicit, not to say enthusiastic, about accepting both.

In order to get clearer about the room left for vital (or mental) forces by the conservation of energy, recall how I earlier distinguished two ways in which early Newtonian theory might allow room for such *sui generis* animate forces. First, such forces might operate spontaneously and indeterministically: nothing in early Newtonian theory would seem to rule out spontaneous forces ungoverned by any deterministic force law. Second, even if the relevant forces are governed by a deterministic force law, they may still be *sui generis*, in the sense that they may be distinct from gravitional forces, chemical forces, and so on, and may arise specifically in living systems or their brains.

The conservation of energy bears differentially on these two kinds of special forces. It does seem inconsistent with the first kind of special force, a spontaneous special force. But it does not directly rule out the second, deterministic kind.

Why should the conservation of energy rule out even a spontaneous special force? (Think of a spontaneous mental force that accelerates molecules in the pineal gland, say.) Why shouldn't such a force simply respect the conservation of energy by not causing accelerations which will violate it? But this doesn't really make sense. The content of the principle of the conservation of energy is that losses of kinetic energy are compensated by buildups of potential energy, and vice versa. But we couldn't really speak of a 'buildup' or 'loss' in the potential energy associated with a force, if there were no force law governing the deployment of that force. So the very idea of potential energy commits us to a law which governs how the relevant force will cause accelerations in the future.

However, nothing in this argument rules out the possibility of vital, mental, or other special forces which *are* governed by deterministic force laws. After all, the conservation of energy in itself does not tell which basic forces operate in the physical universe. Are gravity and impact the only basic forces? What about electromagnetism? Nuclear forces? And so on. Clearly the conservation of energy as such leaves it open exactly which basic forces exist. It requires only that, whatever they are, they operate deterministically and conservatively.[12]

A.5 *Conservative Animism*

In this section I shall briefly sketch the evolution of attitudes to the completeness of physics since Helmholtz's promulgation of the universal conservation of energy. The issues are not straightforward, and there is no question of dealing with them fully here. But I would like to offer at least an outline of how the argument for the completeness of physics has developed since the mid-nineteenth century.

Helmholtz's doctrine left various options open in relation to the completeness of physics. For a start, you could simply deny that the conservation of energy applied to animate forces. That is, you could hold that vital and mental forces are an exception to the general rule that all forces are conservative, and thus insist that the conservation of energy holds only when we are dealing with inanimate forces.

However, this option does not seem to have been popular among

[12] I have the impression that scientifically informed late nineteenth-century philosophers were not particularly exercised by our issue of whether or not there are special vital or mental forces. Understandably enough, they were far more interested in the determinism implied by the conservation of energy even on the assumption of special animate forces. Cf. Tyndall 1898 [1877].

scientifically informed commentators in the second half of the nineteenth century. The doctrine of the universal conservation of energy won widespread acceptance within a decade or two of its formulation. There is of course an evidential question here too: how far was this almost immediate agreement on the conservation of energy dictated by the strength of evidence rather than by intellectual fashion? But there is no question of pursuing this issue here. So let me assume for present purposes that the conservation of energy itself was well supported by the middle of the nineteenth century, and focus instead on where this left the completeness of physics. Certainly this is how the writers I shall discuss henceforth saw the matter. Their question was not whether energy is always conserved, but rather, whether such conservation leaves any room for animate forces.

As I pointed out in the last section, it is clear that conservation does leave such room. The universal conservation of energy may rule out indeterministic animate forces, but there is clearly nothing in it to preclude deterministic animate forces that do respect the conservation of energy. Even so, as I observed, Helmholtz and his young colleagues rejected any such special animate forces. It is interesting to consider what might have persuaded them of this. I suspect that they were moved by what I shall call 'the argument from fundamental forces'. This is the argument that all apparently special forces characteristically *reduce* to a small stock of basic physical forces which conserve energy. Causes of macroscopic accelerations standardly turn out to be composed of a few fundamental physical forces which operate throughout nature. So, while we ordinarily attribute certain physical effects to 'muscular forces', say, or indeed to 'mental causes', we should recognize that these causes, like all causes of physical effects, are ultimately composed of the few basic physical forces.

It is possible that this line of thought was influential in originally persuading Helmholtz of the universal validity of the conservation of energy. We have already seen how Helmholtz's initial formulation of this principle hinged on the assumption that friction and other dissipative forces are non-fundamental forces, macroscopic manifestations of processes involving more fundamental conservative forces. For it is only if we see macroscopic forces like friction as reducing to fundamental conservative forces that we can uphold the universal conservation of energy. Given this view about dissipative forces, a natural move would be to generalize inductively and conclude that *all* apparently special forces must reduce to a small stock of fundamental forces. After all, those special forces which have been quantitatively analysed, like friction, turn out to reduce to more fundamental conservative forces. So this could be seen as providing some inductive reason to conclude that any other apparently special forces, like muscular forces or vital forces or mental forces, will similarly reduce.

Thus consider how Helmholtz argues in *Über die Erhaltung der Kraft*. He takes pains to stress how it is specifically *central* forces independent of time and velocity which ensure the conservation of energy. This emphasis on central forces (by which Helmholtz meant forces which act along the line between the interacting particles) now seems dated. Nowadays conservativeness is normally defined circularly, as a property of those forces which do no work round a closed orbit. This definition does not require a restriction to central forces. However, Helmholtz was in no position to adopt the circular modern definition of conservativeness. He was aiming to *persuade* his readers of the general conservation of energy, so needed an argument. It wouldn't have served simply to observe that energy is conserved by those forces which conserve energy. Helmholtz's actual claim was that energy is conserved by a wide range of known forces: namely, central forces. Still, this by itself doesn't show that energy is conserved by all forces, *unless* all forces are central. Why should this be? Well, as above, one persuasive thought would be that there is a small stock of basic central forces, and that all causes apparently peculiar to special circumstances are composed out of these.

It is clear from our earlier discussion, however, that this reductionist move is not *essential* to a commitment to the universal conservation of energy. An alternative strategy would be to allow that there are *sui generis* animate forces, and to maintain that these fundamental special forces are conservative in their own right. True, this position is open to the objection that there is no direct reason to suppose that any such *sui generis* animate forces *will* be conservative, if they do not reduce to other fundamental conservative forces. But this could be countered with the alternative inductive thought that, since all the *other* fundamental forces so far examined have turned out to be conservative, we should infer that any extra vital or mental fundamental forces will be conservative too.

Somewhat oddly, physiological research in the second half of the nineteenth century added support to this anti-reductionist stance, by offering direct empirical evidence that if there were any special animate forces, they would have to respect the conservation of energy. In a moment I shall argue that physiological research has also given us strong reason to doubt that there *are* any special animate forces. But this latter conclusion derives from investigations at a microscopic cellular level, and such research had to wait until the twentieth century. Prior to that, however, there was a flourishing tradition of energetic research at a more macropscopic level, which identified chemical and energetic inputs and outputs to various parts of the body, and showed that animals are subject to general conservation principles. Especially noteworthy were Max Rubner's elaborate 1889 respiration calorimeter experiments, which showed that the energy emitted by a small dog corresponds exactly to that of the food it consumes. (See Coleman 1971: esp. 140–3.)

The interesting point is that this kind of research did nothing to support the reductionist view that all apparently special forces reduce to a few basic inanimate forces. That normal chemicals are moved around, and that energy is conserved throughout, does not in the end rule out the possibility that some accelerations within bodies are due to special vital or mental forces. It may still be that such forces are activated inside animate creatures, but operate in such a way as to 'pay back' all the energy they 'borrow', and vice versa. Rather, research like Rubner's would have added weight to the position of those who took the existence of *sui generis* animate forces to be consistent with the conservation of energy, as further items in the category of fundamental conservative forces.

As exemplars of this position, I have already mentioned Leibig and Müller, two eminent physiologists of the older generation, who continued to accept vital forces, even after the conservation of energy had won general acceptance. And Brian McLaughlin, in his excellent article on 'British Emergentism' (1992), explains how the philosophers J. S. Mill and Alexander Bain went so far as to argue that the conservation of energy, and in particular the notion of potential energy, lends definite support to the possibility of non-physical forces.[13] (The 'British Emergentists' discussed by McLaughlin constituted a philosophical movement committed precisely to non-physical causes of motion in my sense, causes which were not the vectorial 'resultants' of basic physical forces like gravity and impact, but which 'emerged' when matter arranged itself in special ways. The particular idea which attracted Mill and Bain was that these 'emergent forces' might be stored as unrealized potentials, ready to manifest themselves as causes of motion only when the relevant special circumstances arose.[14])

[13] Indeed this line of thought seems to have become extremely popular in the late nineteenth century. The idea that the brain in a repository of 'nervous energy', which gets channelled in various ways, and is then released in action, is a commonplace of Victorian thinkers from Darwin to Freud.

[14] Not all emergentists were as sophisticated as Mill and Bain. In *Mind and its Place in Nature* (1923), C. D. Broad addresses the issue of whether independent mental causation would violate the conservation of energy (pp. 103–9). But instead of simply claiming that any mental force would operate conservatively, he insists that the principle of the conservation of energy does not explain all motions, even in physical systems, and so leaves room for other causes. He draws

A.6 *The Death of Emergentism*

McLaughlin explains how British Emergentism continued to flourish well into the twentieth century. This highlights the question at issue in this Appendix. Given that thinkers continued to posit special mental and vital forces until well after the Great War, why has the idea of such forces now finally fallen into general disfavour?

Here I think we need to refer to a second line of argument against such forces, an argument from direct physiological evidence. We can view this second argument as operating against the background provided by the earlier argument from fundamental forces. The earlier argument suggested that most natural phenomena, if not all, can be explained by a few fundamental physical forces. This focused the issue of what kind of evidence would demonstrate the existence of extra mental or vital forces. For once we know which other forces exist, then we will know which anomalous accelerations would indicate the presence of special mental or vital forces. Against this background, the argument from physiology is then simply that detailed modern research has failed to uncover any such anomalous physical processes.

As I intimated above, the relevant research dates mostly from the twentieth century. Physiological research in the nineteenth century did not penetrate to the level of forces operating inside bodies. However, in the first half of the twentieth century, the situation changed, and by the 1950s it had become difficult, even for those who were not moved by the abstract reductionist argument from fundamental forces, to continue to uphold special vital or mental forces. A great deal became known about biochemical and neurophysiological processes, especially at the level of the cell, and none of it gave any evidence for the existence of special forces not found elsewhere in nature.

During the first half of the century the catalytic role and protein constitution of enzymes were recognized, basic biochemical cycles were identified, and the structure of proteins analysed, culminating in the discovery of DNA. In the same period, neurophysiological research mapped the body's neuronal network and analysed the electrical mechanisms responsible for neuronal activity. Together, these developments made it difficult to go on maintaining that special forces operate inside living

an analogy with a pendulum on a string, where he says that the 'pull of the string' is a cause which operates independently of any flows of energy, and he suggests that the mind might operate as a similar cause. While it is not entirely clear how Broad intends this analogy to be read, it is difficult to avoid the impression that he has mastered the letter of the conservation of energy, without grasping the wider physical theory in which it is embedded.

bodies. If there were such forces, they could be expected to display some manifestation of their presence. But detailed physiological investigation failed to uncover evidence of anything except familiar physical forces.

In this way, the argument from physiology can be viewed as clinching the case for the completeness of physics against the background provided by the argument from fundamental forces. One virtue of this explanation in terms of these two interrelated arguments is that it yields a natural explanation for the slow advance of the completeness of physics through the century from the 1850s to the 1950s. Imagine a ranking of different thinkers through this period, in terms of the amount of physiological evidence needed to persuade them of completeness, in addition to the abstract argument from fundamental forces. Helmholtz and his colleagues would be at one extreme, in deciding for completeness on the basis of the abstract argument alone, without any physiological evidence. In the middle would be those thinkers who waited for a while, but converted once initial physiological research in the first decades of the twentieth century gave no indication of any forces beyond fundamental forces found throughout nature. At the other end would be those who needed a great deal of negative physiological evidence before giving up on special forces. The existence of this spectrum would thus explain why there was a gradual buildup of support for the completeness of physics as the physiological evidence accumulated, culminating, I would contend, in a general scientific consensus by the 1950s.

Brian McLaughlin offers a rather different explanation for the demise of British Emergentism. He attributes it to the 1920s quantum-mechanical reduction of chemical forces to general physical forces acting on subatomic components (1992: 89). But it seems unlikely that this could have been decisive. After all, why should anybody who was still attracted to *sui generis* animate forces in the 1920s have turned against them simply because of the reduction of *chemistry* to physics? Why should it have mattered to them exactly how many independent forces there were at the level of atoms? At most, the reduction of chemistry to physics would have added some marginal weight to the argument from fundamental forces, by showing that yet another special force reduces to more basic forces. But anybody who had resisted the argument from fundamental forces so far, still upholding vital and mental forces as extra members of the pantheon of fundamental forces into the twentieth century, would surely not be bowled over simply because the physical theorists had now modified the precise inventory of forces operating at the atomic level. To understand why British Emergentism lost ground over the first half of the twentieth century, we need to recognize a different kind of argument: namely, the argument from the emerging findings of physiological research.[15]

A.7 *Conclusion*

This Appendix has charted the history of changing attitudes to the completeness of physics. The important point is that a scientific consensus on completeness was reached only in the middle of the twentieth century. In earlier centuries there was no compelling reason to believe that all physical effects are due to physical causes, and few scientists did believe this. But by the 1950s the combination of the physiological evidence with the argument from fundamental forces left little room for doubt about the doctrine.

In Chapter 1 I raised the question of why philosophical physicalism is peculiarly a creature of the late twentieth century. I hope I have now succeeded in showing that this is no intellectual fad, but a reflection of developments in empirical theory. Without the completeness of physics, there is no compelling reason to identify the mind with the brain. But once the completeness of physics became part of established science, scientifically informed philosophers realized that this crucial premiss could be slotted into a number of variant arguments for physicalism. There seems no

[15] I have presented the conservation of energy as a boundary principle which limits the range of possible forces and helped rule out special animate forces. Barry Loewer has pressed me on whether this emphasis on the conservation of energy is consistent with modern quantum mechanics. Let me make two brief comments. (1) On some interpretations, quantum systems do not always respect the conservation of energy. While energy is conserved in the 'Schrödinger evolution' of quantum systems, it is apparently violated by 'wave collapses'. Some, including myself, take this to argue against wave collapses. But, even if you don't go this way, the conservation of energy will still be respected in Schrödinger evolutions, and this will itself limit the range of (non-collapse-causing) forces. (2) On some, but not all, collapse interpretations, distinctive special causes will be responsible for whether a collapse occurs or not (even though the subsequent chances of the various possible outcomes will still depend entirely on prior physical forces). I am thinking here of interpretations which say that collapses occur when physical systems interact with consciousness (or indeed which say that collapses occur when there are 'measurements', or 'macroscopic interactions', and then refuse to offer any physical explanations of these terms). On these interpretations, the completeness of physics will be violated, as well as the conservation of energy, since collapses don't follow from more basic physical laws, but depend on 'emergent' causes. It would seem an odd victory for anti-materialists, however, if the sole locus of *sui generis* mental action were quantum wave collapses.

reason to look any further to explain the widespread philosophical acceptance of physicalism since the 1950s.

Of course, as with all empirical matters, there is nothing certain here. There is no knock-down argument for the completeness of physics. You could in principle accept the rest of modern physical theory, and yet continue to insist on special mental forces, which operate in as yet undetected ways in the interstices of intelligent brains. And indeed, there do exist bitter-enders of just this kind, who continue to hold out for special mental causes, even after another half-century of ever more detailed molecular biology has been added to the inductive evidence which initially created a scientific consensus on completeness in the 1950s. Perhaps this is what Tyler Burge has in mind when he says that 'materialism is not established, or even deeply supported, by science', or Stephen Clark when he doubts whether anyone could 'rationally suppose' that empirical evidence 'disproves' mind–body dualism. If so, there is no more I can do to persuade them of the completeness of physics. However, I see no virtue in philosophers refusing to accept a premiss which, by any normal inductive standards, has been fully established by over a century of empirical research.

References

Alexander, H. (1956) (ed.), *The Leibniz-Clarke Correspondence* (Manchester: Manchester University Press).

Armstrong, D. (1968), *A Materialist Theory of the Mind* (London: Routledge and Kegan Paul).

Baars, B. (1988), *A Cognitive Theory of Consciousness* (Cambridge: Cambridge University Press).

Balog, K. (1999), 'Conceivability, Possibility, and the Mind-Body Problem'. *Philosophical Review*, 108.

Bigelow, J., and Pargetter, R. (1990), 'Acquaintance with Qualia', *Theoria*, 56.

Block, N. (1978), 'Reductionism: Philosophical Analysis', in W. T. Reich (ed.), *Encyclopaedia of Bioethics* (London: Macmillan).

——(forthcoming), 'The Harder Problem of Consciousness', *Journal of Philosophy*.

——and Stalnaker, R. (2000), 'Conceptual Analysis, Dualism and the Explanatory Gap', *Philosophical Review*, 30.

Broad, C. (1923), *Mind and its Place in Nature* (London: Routledge and Kegan Paul).

Burge, T. (1979), 'Individualism and the Mental', in P. French, T. Uehling and H. Wettstein (eds.), *Studies in Epistemology, Midwest Studies in Philosophy*, vol. 4 (Minneapolis: University of Minnesota Press).

——(1982), 'Other Bodies', in A. Woodfield (ed.), *Thought and Object* (Oxford: Oxford University Press).

——(1993), 'Mind-Body Causation and Explanatory Practice', in J. Heil and A. Mele (eds.), *Mental Causation* (Oxford: Clarendon Press).

Cappelen, H., and Lepore, E. (1997), 'Varieties of Quotation', *Mind*, 106.

Carruthers, P. (2000), *Phenomenal Consciousness* (Cambridge: Cambridge University Press).

Carruthers, P., and Smith, P. (1996) (eds.), *Theories of Theories of Mind* (Cambridge: Cambridge University Press).

Chalmers, D. (1996), *The Conscious Mind* (Oxford: Oxford University Press).

Clark, S. (1996), Review of C. Taliaferro, *Consciousness and the Mind of God*, *Times Literary Supplement*, 23 Feb. 1996.

Coleman, W. (1971), *Biology in the Nineteenth Century* (New York: John Wiley and Sons).

Crane, T. (1991), 'Why Indeed?', *Analysis*, 51.

——(1995), 'The Mental Causation Debate', *Aristotelian Society Supplementary Volume*, 69.

——and Mellor, D. H. (1990), 'There is no Question of Physicalism', *Mind*, 99.

Davidson, D. (1970), 'Mental Events', in L. Foster and J. Swanson (eds.), *Experience and Theory* (London: Duckworth); repr. in Davidson 1980.

——(1980), *Essays on Actions and Events* (Oxford: Clarendon Press).

Davies, M., and Stone, T. (1995*a*) (eds.), *Folk Psychology* (Oxford: Blackwell).

————(1995*b*) (eds.), *Mental Simulation* (Oxford: Blackwell).

Dennett, D. (1978*a*). 'Towards a Cognitive Theory of Consciousness', in *Brainstorms* (Cambridge, Mass.: Bradford Books).

——(1978*b*). 'Why You Can't Make a Computer that Feels Pain', in *Brainstorms* (Cambridge, Mass.: Bradford Books).

——(1991), *Consciousness Explained* (Boston: Little, Brown).

Dretske, F. (1995), *Naturalizing the Mind* (Cambridge, Mass.: MIT Press).

Eilan, N. (1998), 'Perceptual Intentionality, Attention and Consciousness', in A. O'Hear (ed.), *Current Issues in Philosophy of Mind* (Cambridge: Cambridge University Press).

Elkana, Y. (1974), *The Discovery of the Conservation of Energy* (London: Hutchinson).

Faraday, M. (1857), 'On the Conservation of Force', *Proceedings of the Royal Society*, 9.

Feigl, H. (1958), 'The "Mental" and the "Physical"', in H. Feigl, M. Scriven and G. Maxwell (eds.), *Minnesota Studies in the Philosophy of Science*, vol. 2 (Minneapolis: University of Minnesota Press).

Fodor, J. (1990), *A Theory of Content* (Cambridge, Mass.: MIT Press).

Frith, C., Perry, R., and Lumer, E. (1999), 'The Neural Correlates of Cognitive Experience: An Experimental Framework', *Trends in Cognitive Neuroscience*, 3.

Goldstein, H. (1964), *Classical Mechanics*, 2nd edn. (Reading, Mass.: Addison-Wesley).

Goodale, M., and Milner, A. (1992), 'Separate Visual Pathways for Perception and Action', *Trends in Neurosciences*, 15.

Haller, A. Von (1936 [1751]), 'A Dissertation on the Sensible and Irritable Parts of Animals', trans. O. Temkin, *Bulletin of the History of Medicine*, 4.

Harman, G. (1990), 'The Intrinsic Quality of Experience', in J. Tomberlin (ed.), *Philosophical Perspectives*, vol. 4 (Northridge, Calif.: Ridgeview Press).

Helmholtz, H. Von (1847), *Über die Erhaltung der Kraft* (Berlin).

Hill, C. (1997), 'Imaginability, Conceivability, Possibility, and the Mind–Body Problem', *Philosophical Studies*, 87.

——and McLaughlin, B. (1998), 'There are Fewer Things in Reality than are Dreamt of in Chalmers's Philosophy', *Philosophy and Phenomenological Research*, 59.

Horgan, T. (1984), 'Jackson on Physical Information and Qualia', *Philosophical Quarterly*, 34.

Hüttemann, A. (forthcoming), *Micro-Explanation and the Multi-Layered Conception of Reality*.

Jackson, F. (1982), 'Epiphenomenal Qualia', *Philosophical Quarterly*, 32.

——(1986), 'What Mary Didn't Know', *Journal of Philosophy*, 83.

——(1993), 'Armchair Metaphysics', in J. O'Leary-Hawthorne and M. Michael (eds.), *Philosophy in Mind* (Dordrecht: Kluwer).

——(1998), *From Concepts to Metaphysics* (Oxford: Oxford University Press).

Kim, J. (1973), 'Causation, Nomic Subsumption, and the Concept of an Event', *Journal of Philosophy*, 70.

——(1998), *Mind in a Physical World* (Cambridge, Mass.: MIT Press).

Kripke, S. (1971), 'Identity and Necessity', in M. Munitz (ed.), *Identity and Individuation* (New York: New York University Press).

——(1972), 'Naming and Necessity', in D. Davidson and G. Harman (eds.), *Semantics of Natural Languages* (Dordrecht: Reidel).

——(1980), *Naming and Necessity* (Cambridge, Mass.: Harvard University Press).

Leibniz, G. (1898 [1696]), *The Monadology*, trans. R. Latta (Oxford).

Levine, J. (1983), 'Materialism and Qualia: The Explanatory Gap', *Pacific Philosophical Quarterly*, 64.

——(1993), 'On Leaving Out What It's Like', in M. Davies and G. Humphreys (eds.), *Consciousness: Psychological and Philosophical Essays* (Oxford: Blackwell).

Lewis, D. (1966), 'An Argument for the Identity Theory', *Journal of Philosophy*, 63.

Lewis, D. (1983), 'Postscript to "Mad Pain and Martian Pain"', in his *Philosophical Papers*, vol. 1 (Oxford: Blackwell).

——(1988), 'What Experience Teaches', *Proceedings of the Russellian Society of Sydney University*; reprinted in W. Lycan (ed.), *Mind and Cognition: A Reader* (Oxford: Blackwell, 1990).

Libet, B. (1993), 'The Neural Time Factor in Conscious and Unconscious Events', in *Experimental and Theoretical Studies of Consciousness* (Ciba Foundation Symposium, 174; London: John Wiley and Sons).

Loar, B. (1990), 'Phenomenal States', in J. Tomberlin (ed.), *Philosophical Perspectives*, vol. 4; repr. in a revised form in N. Block, O. Flanagan, and G. Guzeldere (eds.), *The Nature of Consciousness* (Cambridge, Mass.: MIT Press, 1997).

——(1999), 'Should the Explanatory Gap Perplex Us?', in T. Rockmore (ed.), *Proceedings of the Twentieth World Congress of Philosophy*, vol. 2 (Bowling Green, Oh.: Philosophy Documentation Center).

Lockwood, M. (1989), *Mind, Brain and Quantum* (Oxford: Blackwell).

Lycan, W. (1987), *Consciousness* (Cambridge, Mass.: MIT Press).

——(1996), *Consciousness and Experience* (Cambridge, Mass.: MIT Press).

McGinn, C. (1991), *The Problem of Consciousness* (Oxford: Blackwell).

Mack, A., and Rock, I. (1998), *Inattentional Blindness* (Cambridge, Mass.: MIT Press).

McLaughlin, B. (1992), 'The Rise and Fall of British Emergentism', in A. Beckerman, H. Flohr, and J. Kim (eds.), *Emergence or Reduction* (New York: De Gruyter).

Martin, M. (forthcoming), 'The Transparency of Experience'.

Mellor, D. H. (1995), *The Facts of Causation* (London: Routledge).

Millikan, R. (1984), *Language, Thought, and Other Biological Categories* (Cambridge, Mass.: MIT Press).

——(1989), 'Biosemantics', *Journal of Philosophy*, 86.

——(1993), 'White Queen Psychology', in *White Queen Psychology and Other Essays for Alice* (Cambridge, Mass.: MIT Press).

——(2000), *On Clear and Confused Ideas* (Cambridge: Cambridge University Press).

Nagel, T. (1974), 'What is it Like to be a Bat?' *Philosophical Review*, 83.

Neander, K. (1998), 'The Division of Phenomenal Labour', in J. Tomberlin (ed.), *Philosophical Perspectives*, vol. 12 (Northridge, Calif.: Ridgeview Press).

Nemirow, L. (1990), 'Physicalism and the Cognitive Role of Acquaintance', in W. Lycan (ed.), *Mind and Cognition: A Reader* (Oxford: Blackwell).

Newton, I. (1952 [1704]), *Opticks* (New York: Dover).

——(1966 [1686]), *Mathematical Principles of Natural Philosophy*, ed. F. Cajori (Berkeley: University of California Press).

Nisbett, R., and Ross, L. (1980), *Human Inference* (New York: Prentice-Hall).

——and Wilson, T. (1977), 'Telling More Than We Can Know', *Psychological Review*, 84.

Oppenheim, H., and Putnam, P. (1958), 'Unity of Science as a Working Hypothesis', in H. Feigl, M. Scriven, and G. Maxwell (eds.), *Concepts, Theories and the Mind–Body Problem*, Minnesota Studies in the Philosophy of Science, vol. 2 (Minneapolis: University of Minnesota Press).

Papineau, D. (1977), 'The *Vis Viva* Controversy: Do Meanings Matter?', *Studies in History and Philosophy of Science*, 8.

——(1984), 'Representation and Explanation', *Philosophy of Science*, 51.

——(1993*a*), *Philosophical Naturalism* (Oxford: Blackwell).

——(1993*b*). 'Physicalism, Consciousness, and the Antipathetic Fallacy', *Australasian Journal of Philosophy*, 71.

——(1995), 'The Antipathetic Fallacy and the Boundaries of Consciousness', in T. Metzinger (ed.), *Conscious Experience* (Paderborn: Ferdinand Schoningh).

——(1996), 'Theory-Dependent Terms', *Philosophy of Science*, 63.

——(1998), 'Mind the Gap', in J. Tomberlin (ed.), *Philosophical Perspectives*, vol. 12 (Northridge, Calif.: Ridgeview Press).

——(1999), 'Normativity and Judgement', *Aristotelian Society Supplementary Volume*, 73.

——(2000), 'The Rise of Physicalism', in M. Stone, and J. Wolff (eds.), *The Proper Ambition of Science* (London: Routledge); and in C. Gillett and B. Loewer (eds.), *Physicalism and its Discontents* (New York: Cambridge University Press, forthcoming).

——and Spurrett, D. (1999), 'A Note on the Completeness of "Physics" ', *Analysis*, 59.

Peacocke, C. (1989), 'No Resting Place: A Critical Notice of *The View from Nowhere*', *Philosophical Review*, 98.

Penrose, R. (1994), *Shadows of the Mind* (Oxford: Oxford University Press).

Place, U. T. (1956), 'Is Consciousness a Brain Process?', *British Journal of Psychology*, 47.

Putnam, H. (1975), 'The Meaning of "Meaning" ', in K. Gunderson (ed.), *Language, Mind, and Knowledge*, Minnesota Studies in the Philosophy of Science, vol. 7 (Minneapolis: University of Minnesota Press).

Rey, G. (1991), 'Sensations in a Language of Thought', in E. Villeneuva (ed.), *Philosophical Issues*, vol. 2 (Northridge, Calif.: Ridgeview).

Rosenthal, D. (1986), 'Two Concepts of Consciousness', *Philosophical Studies*, 49.

Russell, B. (1927), *The Analysis of Matter* (London: Kegan Paul).

Saka, P. (1998), 'Quotation and the Use–Mention Distinction', *Mind*, 107.

Schiffer, S. (1987), *Remnants of Meaning* (Cambridge, Mass.: MIT Press).

Segal, G. (2000), *A Slim Book about Narrow Content* (Cambridge, Mass.: MIT Press).

——and Sober, E. (1991), 'The Causal Efficacy of Content', *Philosophical Studies*, 63.

Shallice, T. (1988), *From Neuropsychology to Mental Structure* (Cambridge: Cambridge University Press).

Smart, J. J. C. (1959), 'Sensations and Brain Processes', *Philosophical Review*, 68.

Smith, C. (1998), *The Science of Energy* (London: Athlone Press).

Spelke, E. (1991), 'Physical Knowledge in Infancy', in S. Carey and R. Gelman (eds.), *The Epigenesis of Mind* (Hillsdale, NJ: Erlbaum).

Steigerwald, J. (1998), '*Lebenskraft* in Reflection' (Ph.D. dissertation, University of London).

Sturgeon, S. (1994), 'The Epistemic View of Subjectivity', *Journal of Philosophy*, 91.

——(1998), 'Physicalism and Overdetermination', *Mind*, 107.

——(2000), *Matters of Mind* (London: Routledge).

Tye, M. (1995), *Ten Problems of Consciousness* (Cambridge, Mass.: MIT Press).

——(1999), 'Phenomenal Consciousness: The Explanatory Gap', *Mind*, 108.

——(2000), *Consciousness, Color and Content* (Cambridge, Mass.: MIT Press).

Tyndall, J. (1898 [1877]). 'Science and Man', in *Fragments of Science*, vol. 2 (New York).

Weiskrantz, L. (1986), *Blindsight* (Oxford: Oxford University Press).

Whytt, R. (1755), *Physiological Essays* (Edinburgh).

Wilson, T., Hull, J., and Johnson, J. (1981), 'Awareness and Self-Perception: Verbal Reports on Internal States', *Journal of Personality and Social Psychology*, 40.

Witmer, G. (2000), 'Locating the Overdetermination Problem', *British Journal for the Philosophy of Science*, 51.

Wittgenstein, L. (1953), *Philosophical Investigations* (Oxford: Blackwell).

Woolhouse, R. (1985), 'Leibniz's Reaction to Cartesian Interaction', *Procedings of the Aristotelian Society*, 86.

Yablo, S. (1992), 'Mental Causation', *Philosophical Review*, 101.

Index

Printed in the United States
19669LVS00001B/36